R. Johnson

August, 1989.

MONOGRAPHS ON
STATISTICS AND APPLIED PROBABILITY

General Editors

D.R. Cox, D.V. Hinkley, D. Rubin and B.W. Silverman

1 Stochastic Population Models in Ecology and Epidemiology
M.S. Bartlett (1960)

2 Queues *D.R. Cox and W.L. Smith* (1961)

3 Monte Carlo Methods *J.M. Hammersley and D.C. Handscomb* (1964)

4 The Statistical Analysis of Series of Events *D.R. Cox and
P.A.W. Lewis* (1966)

5 Population Genetics *W.J. Ewens* (1969)

6 Probability, Statistics and Time *M.S. Bartlett* (1975)

7 Statistical Inference *S.D. Silvey* (1975)

8 The Analysis of Contingency Tables *B.S. Everitt* (1977)

9 Multivariate Analysis in Behavioural Research *A.E. Maxwell* (1977)

10 Stochastic Abundance Models *S. Engen* (1978)

11 Some Basic Theory for Statistical Inference *E.J.G. Pitman* (1979)

12 Point Processes *D.R. Cox and V. Isham* (1980)

13 Identification of Outliers *D.M. Hawkins* (1980)

14 Optimal Design *S.D. Silvey* (1980)

15 Finite Mixture Distributions *B.S. Everitt and D.J. Hand* (1981)

16 Classification *A.D. Gordon* (1981)

17 Distribution-free Statistical Methods *J.S. Maritz* (1981)

18 Residuals and Influence in Regression *R.D. Cook and S. Weisberg* (1982)

19 Applications of Queueing Theory *G.F. Newell* (1982)

20 Risk Theory, 3rd edition *R.E. Beard, T. Pentikäinen and
E. Pesonen* (1984)

(Full details concerning this series are available from the publishers)

Analysis of Binary Data

SECOND EDITION

D.R. COX

Nuffield College,
Oxford

and

E.J. SNELL

Department of Mathematics,
Imperial College London

LONDON   NEW YORK
CHAPMAN AND HALL

First published in 1970 by Methuen & Co. Ltd
Reprinted 1977 by Chapman and Hall Ltd
11 New Fetter Lane, London EC4P 4EE
Reprinted 1980, 1983, 1987

Published in the USA by Chapman and Hall
29 West 35th Street, New York NY 10001

Second edition 1989

© 1970 D.R. Cox, 1989 D.R. Cox and E.J. Snell

Typeset in 10/12 Times by Thomson Press (India) Ltd, New Delhi

Printed in Great Britain by St Edmundsbury Press Ltd
Bury St Edmunds, Suffolk

ISBN 0 412 30620 4

British Library Cataloguing in Publication Data

Cox, D.R. (David Roxbee), 1924–
   Analysis of binary data.—2nd ed.
   1. Probabilities & statistical mathematics
   I. Title   II. Snell, E.J.   III. Series
   519.2
   ISBN 0-412-30620-4

Library of Congress Cataloging in Publication Data

Cox, D.R. (David Roxbee)
  Analysis of binary data/D.R. Cox and E.J. Snell.—2nd ed.
    p.   cm.—(Monographs on statistics and applied probability)
  Includes bibliographies and index.
  ISBN 0-412-30620-4
  1. Analysis of variance.   2. Probabilities.   3. Distribution
  (Probability theory)   I. Snell, E.J.   II. Title.   III. Series.
  QA279.C68   1989
  519.5'352—dc19

Contents

Preface to first edition

This monograph concerns the analysis of binary (or quantal) data, i.e. data in which an observation takes one of two possible forms, e.g. success or failure. The central problem is to study how the probability of success depends on explanatory variables and groupings of the material.

Many particular methods, especially significance tests, have been proposed for such problems and one of the main themes of the monograph is that these methods are unified by considering models in which the logistic transform of the probability of success is a linear combination of unknown parameters. These linear logistic models play here much the same role as do normal-theory linear models in the analysis of continuously distributed data.

Some knowledge of the theory of statistics is assumed. I have written primarily for statisticians, but I hope also that scientists and technologists interested in applying statistical methods will, by concentrating on the examples, find something useful here.

I am very grateful to Dr Agnes M. Herzberg and Dr P.A.W. Lewis for extremely helpful comments. I acknowledge also the help of Mrs Jane Gentleman who programmed some of the calculations.

D.R. Cox
London
April 1969

Preface to second edition

We have added new material partly to amplify matters dealt with only very cryptically in the first edition and partly to describe some of the more recent developments, for example on regression diagnostics. In addition the contents of the first edition have been rearranged; for example, the method of least squares with empirically estimated weights is now of much less importance than it used to be because computational developments have, for many problems, brought maximum likelihood fitting within the painless grasp of most users of statistical analysis. By giving some prominence to examples we have aimed to make the book accessible to a range of readers. One of the Appendices summarizes the theoretical background.

When the first edition was written it was feasible to give a relatively complete annotated bibliography of work on the analysis of binary data. The number of papers on this topic is now so large and is increasing so rapidly that no such bibliography has been attempted. Instead, in the Bibliographic Notes at the end of the each chapter we have aimed to give just a few key references for further reading and for details omitted from the main text.

We are grateful to Professor N. Wermuth, Mainz, for thoughtful comments on a portion of the manuscript and to Professor J.K. Lindsey, Liége, for advice over original data.

The first author is grateful to Science and Engineering Research Council for a Senior Research Fellowship held at Department of Mathematics, Imperial College, London.

D.R. Cox
E.J. Snell
London
July 1988

CHAPTER 1

Binary response variables

1.1 Introduction

Suppose that on each individual we have an observation that takes one of two possible forms. The following are examples:

1. an electronic component may be defective, or may be nondefective;
2. a test animal may die from a specified dose of a poison, or may survive;
3. a subject may give the correct reply in an experimental situation, or may give a wrong reply;
4. a test specimen may fracture when struck with a standardized blow, or may not;

and so on. If for the ith individual we can represent this observation, or response, by a random variable, Y_i, we may without loss of generality code the two possible values of Y_i by 1 and 0 and write

$$E(Y_i) = \operatorname{prob}(Y_i = 1) = \theta_i, \quad \operatorname{prob}(Y_i = 0) = 1 - \theta_i, \quad (1.1)$$

say. It is often convenient to call $Y_i = 1$ a 'success' and $Y_i = 0$ a 'failure'. It is reasonable to call such observations binary; an older term is quantal.

We assume that such binary observations are available on n individuals, usually assumed to be independent. The problem is to develop good methods of analysis for assessing any dependence of θ_i on explanatory variables representing, for example, groupings of the individuals or quantitative explanatory variables.

We have followed the usual terminology and have distinguished between (a) response variables and (b) explanatory variables, the variables of the second type being used to explain or predict variation in variables of the first type.

Sometimes a binary response variable arises by condensing a more complex response. Thus a component may be classed as defective when a quantitative test observation falls outside specification limits or, more generally, when a set of test observations falls in an unacceptable region. When this is done we need to consider whether there is likely to be serious loss in treating the problem in terms of a binary response.

In addition to a binary response variable there may be further response variables for each individual. Thus in a psychological experiment, as well as the rightness or wrongness of a reply, the time taken to make the reply may be available. Joint analysis of the response variables is then likely to be informative.

1.2 Examples

It is convenient to begin with a few simple specific examples illustrating the problems to be considered.

Example 1.1 The 2 × 2 contingency table
Suppose that there are two groups of individuals, 0 and 1, of sizes n_0 and n_1 and that on each individual a binary response is obtained. The groups may, for example, correspond to two treatments. Suppose further that we provisionally base the analysis on the assumption that all individuals respond independently with probability of success depending only on the group, and equal, say, to ϕ_0 and ϕ_1 in the two groups. Note that we use ϕ for a probability referring to a group, reserving θ for the probability of success for an individual.

In this situation we need consider only the random numbers of successes R_0 and R_1 in the two groups; in fact (R_0, R_1) form sufficient statistics for the unknown parameters (ϕ_0, ϕ_1). It is conventional to set out the numbers of successes and failures in a 2 × 2 contingency table as in Table 1.1. Nearly always it is helpful to calculate the proportions of successes in the two groups.

In Table 1.1 the columns refer to a dichotomy of the response variable, i.e. to a random variable in the mathematical model; the rows refer to a factor classifying the individuals into two groups, the classification being considered as non-random for the purpose of the analysis. It is possible to have contingency tables in which both rows and columns correspond to random variables.

As a specific example, the data of Table 1.2 refer to a retrospective survey of physicians (Cornfield, 1956). Data were obtained on a group of lung cancer patients and a comparable control group. The numbers

Table 1.1 *A 2 × 2 contingency table*

	Failures	Successes	Total	Propn successes
Group 0	$n_0 - R_0$	R_0	n_0	R_0/n_0
Group 1	$n_1 - R_1$	R_1	n_1	R_1/n_1
Total	$n_0 + n_1 - R_0 - R_1$	$R_0 + R_1$	$n_0 + n_1$	

Table 1.2 *Numbers of smokers in two groups of physicians*

	Smokers	Non-smokers	Total	Propn non-smokers
Controls	32	11	43	0.256
Lung cancer patients	60	3	63	0.048
Total	92	14	106	

of individuals in the two groups are approximately equal and are in no way representative of population frequencies. Hence it is reasonable to make an analysis conditionally on the observed total numbers in the two groups.

An essential initial step is to calculate the sample proportions of successes in the two groups R_0/n_0 and R_1/n_1, and these are shown in Table 1.2. Further analysis of the table is concerned with the precision of these proportions.

When two groups are to be compared using binary observations, it will often be sensible to make an initial analysis from a 2×2 contingency table. However, the assumptions required to justify condensation of the data into such a form are not to be taken lightly. Thus in Section 2.4 we shall deal with the methods to be followed when pairs of individuals in the two groups are correlated. The most frequent inadequacy of an analysis by a single 2×2 contingency table is, however, the presence of further factors influencing the response, i.e. nonconstancy of the probability of success within groups. To ignore such further factors can be very misleading.

For many purposes it is not necessary to give the numbers of failures in addition to the numbers of successes and the numbers of trials per group. In most of the more complex examples we shall therefore omit the first column of the 2×2 table.

There are many extensions of the 2×2 table; there follow some examples.

Example 1.2 Several 2×2 contingency tables
Suppose that to compare two treatments we have several sets of observations, each of the form of Example 1.1. The different sets may correspond to levels of a further factor or, as in the following specific example, may correspond to different blocks of an experimental design. In general, in the sth set of data, let $n_{0,s}$ and $n_{1,s}$ be the sample sizes in the two groups and let $R_{0,s}$ and $R_{1,s}$ be the total numbers of successes.

Table 1.3 (Gordon and Foss, 1966) illustrates this. On each of 18 days babies not crying at a specified time in a hospital ward served as subjects. On each day one baby chosen at random formed the experimental group and the remainder were controls. The binary response was whether the baby was crying or not at the end of a specified period. In Table 1.3, not crying is taken as a 'success' and the observed numbers $r_{0,s}$ and $r_{1,s}$ are therefore the numbers of babies in the two groups not crying; the common convention is

Table 1.3 *The crying of babies*

Day, s	No of control babies, $n_{0,s}$	No. not crying, $r_{0,s}$	No. of experimental babies, $n_{1,s}$	No. not crying, $r_{1,s}$
1	8	3	1	1
2	6	2	1	1
3	5	1	1	1
4	6	1	1	0
5	5	4	1	1
6	9	4	1	1
7	8	5	1	1
8	8	4	1	1
9	5	3	1	1
10	9	8	1	0
11	6	5	1	1
12	9	8	1	1
13	8	5	1	1
14	5	4	1	1
15	6	4	1	1
16	8	7	1	1
17	6	4	1	0
18	8	5	1	1

followed of denoting observed values of random variables by lower-case letters. The special feature of this example is that $n_{1,s} = 1$, so that $R_{1,s}$ takes values 0 and 1; usually there are several individuals in each group.

The object of the analysis is to assess the effect of the treatment on the probability of success. The tentative basis for the analysis is that there is in some sense a constant treatment effect throughout the experiment, even though there may be some systematic variation from day to day. The experiment has the form of a randomized block design, in fact a matched pair design, but the binary nature of the response and the varying numbers of individuals in the groups complicate the analysis.

For the reasons indicated after Example 1.1 it would not in general be a sound method of analysis to pool the data over days, thus forming a single 2×2 contingency table with entries $\Sigma R_{0,s}, \Sigma R_{1,s}$, etc.

One simple, if approximate, method of analysis that is not distorted by systematic differences between groups is to calculate for the sth group the difference in the proportions of successes, i.e.

$$\frac{R_{1,s}}{n_{1,s}} - \frac{R_{0,s}}{n_{0,s}}. \tag{1.2}$$

This is an unbiased estimate for the sth set of the difference between the probabilities of success. When (1.2) is averaged over all the sets, an unbiased estimate of the mean difference between groups results. A difficulty of this analysis is that the quantities (1.2) have in general different precisions for different s.

Later, further examples will be given whose analysis requires the combination of data from several 2×2 contingency tables.

Example 1.3 A 2×2^p system

Suppose that there are p two-level factors thought to affect the probability of success. Let a binary response be observed on each individual and suppose that there is at least one individual corresponding to each of the 2^p cells, i.e. possible factor combinations; usually there are an appreciable number of individuals in each cell.

It would be possible to think of such data as arranged in a $(p + 1)$-dimensional table, with two levels in each dimension. Alternatively and more usefully, we can think of a 2×2^p table in which the two columns correspond to success and failure and the 2^p rows are the 2^p standard treatments of the factorial system. One of the problems of

Table 1.4 *2 × 2⁴ system. Study of cancer knowledge*

	No. successes	No. trials	Propn successes		No. successes	No. trials	Propn successes
1	84	477	0.176	*d*	2	12	0.167
a	75	231	0.325	*ad*	7	13	0.538
b	13	63	0.206	*bd*	4	7	0.571
ab	35	94	0.372	*abd*	8	12	0.667
c	67	150	0.447	*cd*	3	11	0.273
ac	201	378	0.532	*acd*	27	45	0.600
bc	16	32	0.500	*bcd*	1	4	0.250
abc	102	169	0.604	*abcd*	23	31	0.742

analysis is to examine what order of interactions, in some suitable sense, is needed to represent the data; that is, we think first of 2^p probabilities of success, one for each cell, and then try to represent these usefully in terms of a smaller number of parameters.

Table 1.4 is a specific example of a 2×2^4 system based on an observational study by Lombard and Doering (1947) (see Dyke and Patterson (1952) for a detailed discussion of the analysis of these data). In this study, the response concerned individuals' knowledge of cancer, as measured in a test, a 'good' score being a success and a 'bad' score a failure. There were four factors expected to account for variation in the probability of success, the individuals being classified into 2^4 cells depending on presence or absence of exposure to A, newspapers; B, radio; C, solid reading; D, lectures. In Table 1.4 the standard notation for factor combinations in factorial experiments is used; thus *ac* denotes the cell in which A and C are at their upper levels and B and D at their lower levels.

Some general conclusions can be drawn from inspection of the cell proportions of successes. This is an example where both the response variable and the factors are reduced to two levels from a more complex form.

Example 1.4 Serial order

Suppose that a series of independent binary responses is observed and that it is suspected that the probability of success changes systematically with serial order. One application is to some types of data in experimental psychology, where a subject makes a series of responses, each either correct or incorrect, and where the probability of correct response is suspected of changing systematically. In this context, the treating of successive responses as independent may, however, be seriously misleading. Another application is in human genetics where each child in a family is classified as having or not having a particular genetic defect. Here, except for possible complications from multiple births, each family leads to a sequence of binary responses. In this application, data from many families are required. We then have a number of sequences, usually not all of the same length, and it is required to examine the data for systematic changes, with serial order, in the probability of a genetic defect. It is not usually reasonable to suppose that in the absence of such effects the probability of a defect is the same for all families.

Example 1.5 Stimulus binary response relation
The following situation is of wide occurrence. There is a stimulus under the experimenter's control; each individual is assigned a level of the stimulus and a binary response then observed. One important field of application is bioassay, where, for example, different levels of stimulus may represent different doses of a poison, and the binary response is death or survival. Similar situations arise in many other fields.

In such applications it is often possible to choose a measure x of stimulus level such that the probability of success is zero for large negative x, unity for large positive x and is a strictly increasing function of x. In fact it has the mathematical properties of a continuous cumulative distribution function; see Fig. 1.1(a). If the x scale is suitably chosen, the distribution function will be symmetric; for example, in the particular application mentioned above it is often helpful to take x as log dose.

Table 1.5 gives some illustrative data; at each of a number of dose levels a group of individuals is tested and the number dying recorded. In Fig. 1.1(b) the proportions dying are plotted against log

Table 1.5 *Simple form of bioassay*

Concn	Log_2 concn	No. of deaths	No. of indivs	Propn deaths
c_0	0	2	30	0.067
$2c_0$	1	8	30	0.267
$4c_0$	2	15	30	0.500
$8c_0$	3	23	30	0.767
$16c_0$	4	27	30	0.900

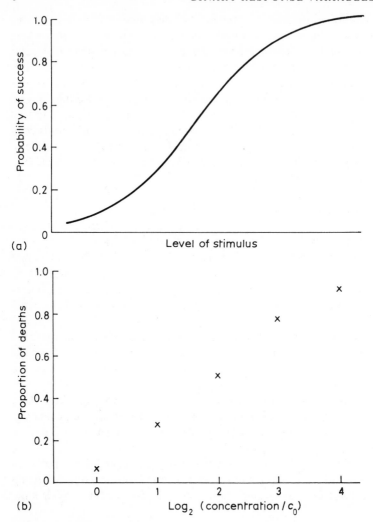

Figure 1.1 *Stimulus–response curves. (a) Idealized theoretical curve, (b) empirical curve from Table 1.5.*

concentration and, except for random fluctuations, give a curve similar to the idealized theoretical one of Fig. 1.1(a). The object of analysing such data is to summarize and assess the properties of the stimulus response curve. Sometimes, as in many bioassay problems, the aspect of primary importance is the level of x at which the probability of success is $\frac{1}{2}$, or sometimes the level x_p at which some other specified value, p, of the probability of success is achieved. In other applications, for example in experimental psychology, the steepness of the response curve is the aspect of primary interest. We shall refer to x variously as a stimulus, or as an explanatory variable.

Example 1.6 Grouped data with binary response
Suppose that there are $k = g + 1$ groups of individuals, each individual having a binary response. If individuals' responses are independent, and the probability of success is constant within each group, we can, just as in the discussion of the 2×2 table, condense the data, giving simply the number of trials n_s and the number of successes R_s for the sth group $(s = 0, \ldots, g)$. In contingency table form we then have Table 1.6.

Now suppose that the groups are meaningfully ordered and that it is reasonable to expect that any change in the probability of success is monotonic with group order. If, further, scores $x_0, \ldots, x_g$ can be allocated to the groups such that a smooth relation between the probability of success and the value of x is reasonable, the situation is

Table 1.6 *A contingency table for grouped data*

	Failures	Successes	Total	Propn successes
Group 0	$n_0 - R_0$	R_0	n_0	R_0/n_0
$\vdots$	$\vdots$	$\vdots$	$\vdots$	$\vdots$
Group g	$n_g - R_g$	R_g	n_g	R_g/n_g

Table 1.7 *Nasal carrier rate and tonsil size*

	Non-carriers	Carriers	Total	Propn of carriers	x
Tonsils present, not enlarged	478	19	497	0.0382	-1
Tonsils enlarged +	531	29	560	0.0492	0
Tonsils enlarged + +	270	23	293	0.0819	1

then formally the same as that of Example 1.5. A distinction is that in the previous example the main interest is in the location and shape of a response curve, whereas in the present example there is usually substantial interest in the null hypothesis that the probability of success is constant.

Table 1.7 gives a specific example quoted by Armitage (1955). The data refer to children aged 0–15 and the binary response concerns whether a child is or is not a carrier for *Streptococcus pyogenes*; the children are grouped into three sets depending on tonsil size. No objective x variable was available and the tentative scores $-1, 0, 1$ were therefore assigned to the three groups. A difficulty in the interpretation of such data is that other possibly relevant explanatory variables – age, sex, etc. – should be considered.

Example 1.7 Multiple regression with binary response
The previous three examples have involved a dependence between a probability of success and a single regressor or explanatory variable, x; they are thus comparable with normal-theory regression problems with a single regressor variable. Suppose now that there are for the ith individual, p, explanatory variables $x_{i1}, \ldots, x_{ip}$, regarded as non-random, and a binary response. It is necessary to assess the relation between the probability of success and the variables $x_{i1}, \ldots, x_{ip}$.

This type of problem occurs especially in medical contexts. The binary response may represent success or failure of a particular treatment; death or survival over a specified time following treatment; or death from a particular cause, as contrasted with death from some other cause. The variables $x_{i1}, \ldots, x_{ip}$ represents quantitative, or qualitative, properties of an individual thought to influence the response; possible x variables are age, time since diagnosis, initial severity of symptoms, sex (scored as a zero–one variable), aspects of medical history of the individual, etc.

One particular example is a study of the factors affecting the probability of having coronary heart disease (see Walker and Duncan (1967) who give a number of earlier references). Another example concerns perinatal mortality (Feldstein, 1966).

Example 1.8 Regression on two explanatory variables
Table 1.8 summarizes a two-factor 5×4 industrial investigation in which the number, r, of ingots not ready for rolling out of m tested is

pp. 11, 36, 39-41, 66

Table 1.8 *Number, r, of ingots not ready for rolling out of m tested*
(First figure in each cell is *r*, second *m*)

Soaking time, x_2	Heating time, x_1				
	7	14	27	51	Total
1.0	0, 10	0, 31	1, 56	3, 13	4, 110
1.7	0, 17	0, 43	4, 44	0, 1	4, 105
2.2	0, 7	2, 33	0, 21	0, 1	2, 62
2.8	0, 12	0, 31	1, 22	0, 0	1, 65
4.0	0, 9	0, 19	1, 16	0, 1	1, 45
Total	0, 55	2, 157	7, 159	3, 16	12, 387

shown for combinations of heating time, x_1, and soaking time, x_2. This
is a rather simpler situation than most of those sketched in
Example 1.7 because there are only two explanatory variables. Note
also that the data are grouped into sets all with the same values of the
explanatory variables.

In some situations, especially with observational data, we need
to analyse changes in the probability of success, there being a
considerable number of potential explanatory variables. Two broad
approaches are possible, illustrated by Examples 1.3 and 1.7.

In the first approach the values of the explanatory variables are
coarsely grouped, in the extreme case each explanatory variable
taking only two values, as in Example 1.3. With *p* explanatory
variables there will thus be at least 2^p cells, for each of which the
proportion of successes can be found. An advantage of this approach
is that quite complicated 'interactions' can be detected. Possible
disadvantages stem from the necessity of coarse grouping and from
the fact that if *p* is at all large many of the cells will either be empty or
contain very few observations.

In the second approach, a regression-like model is taken, express-
ing a smooth and simple dependence of the probability of success on
the values of the explanatory variables. No grouping is necessary, but
a disadvantage is that relatively complicated interactions may be
difficult to detect. In practice both methods are useful, separately and
in combination. With both, problems concerning alternative choices

of explanatory variables, familiar from normal-theory regression methods, arise in essentially the same form.

Example 1.9 A binary time series
In the previous examples it is a reasonable provisional assumption to suppose that the responses of different individuals are independent. In the analysis of a binary time series we are directly concerned with the lack of independence of different observations.

A specific example concerns daily rainfall. It is often reasonable first to classify days as wet (success) or dry (failure); there results a sequence of 1s and 0s, a binary time series. The amounts of rainfall on wet days can be analysed separately.

More generally, if we consider a response 1 as the occurrence of an event and a response 0 as non-occurrence, a binary time series is a series of events in discrete time. If, further, the proportion of 1s is low, the series approximates to a series of point events in continuous time; in particular, a completely random binary series, in which all responses have independently the same probability, θ, of giving a success, tends, as $\theta \to 0$, to a Poisson process. Cox and Lewis (1966) have summarized statistical techniques for the analysis of point events in continuous time.

In this section a number of relatively simple problems have been described which can be generalized in various ways:

1. we may have situations of more complex structure. Thus, instead of the single response curve of Example 1.5, we might have several response curves and be interested in comparing their shapes;
2. we may have multivariate binary responses and consider problems analogous to those of multivariate normal theory;
3. we may have responses taking not just two possible values, but some small number greater than two. Can the techniques to be developed for analysing binary responses be extended?

Some of the more complex problems will be considered later. Others are indicated as further results or exercises.

As in other fields of statistical analysis, problems of two broad types arise. We require techniques for efficient analysis and assessment of uncertainty in the context of an assumed probabilistic model. Also we need techniques for tabular and graphical display and condensation of data, sometimes with the objective of finding a suitable model for more detailed analysis. This aspect is relatively more important with extensive data. This book deals mainly but by no means entirely with the first type of technique.

1.3 Dependency relations for probabilities

The primary theme of this book is the study of the dependence of $\theta = \text{prob}(Y = 1)$, the probability of 'success' of a binary response variable, Y, on a vector x of explanatory variables, including groupings of the data. Therefore a key issue is how such dependencies are to be specified mathematically.

There are broadly three approaches. The first is to use fairly simple empirical functions that express θ in terms of x and some unknown parameters in a reasonably flexible way, such that the parameters have a clear interpretation and preferably such that the resulting statistical analysis is straightforward. The second approach is to relate problems for binary responses to those for some underlying latent (i.e. not directly observable) continuous response variable. Unless the latent variable is in principle observable via some other type of investigation this approach is best regarded as a device for discovering suitable functions for use in the first approach. The third route is to look for smooth relations, usually not specified parametrically, uncovered essentially by merging information from individuals whose explanatory variables are close together in some sense. We shall not discuss this approach in detail.

At first sight the simplest empirical relation is to suppose that θ_i, the value of θ for the ith individual, is linearly related to the explanatory variables, i.e. for the ith individual

$$\theta_i = \alpha + x_i\beta = \alpha + \sum x_{is}\beta_s, \qquad (1.3)$$

where x_i is a row vector of explanatory variables for the ith individual, β is a column vector of unknown regression coefficients and α is an unknown intercept. Note that in general theoretical discussions it is sensible to omit the separate α and to incorporate it into the vector β by making one component explanatory variable a vector of 1s. We shall not do this in this initial discussion.

Equation (1.3) is exactly the standard sort of linear model used for quantitative response variables; this makes it tempting to treat the binary responses as if they were quantitative, i.e. to score them 0 or

1, and to apply ordinary regression and analysis of variance methods; we comment further on this below.

The most serious restriction on the usefulness of (1.3) arises from the condition

$$0 \leqslant \theta_i \leqslant 1. \tag{1.4}$$

This may cause some difficulties in fitting, especially if ordinary least squares is used, but there is the more major matter that the parameters in the model (1.3) inevitably have a limited interpretation and range of validity. Consider for instance the analysis of the stimulus–response curve of Example 1.5 and suppose that we take a model (1.3) in which the probability of success is a linear function of the variable x. Even if the data were satisfactorily linear over the observed range of stimuli, it is certain that the linear relation will fail outside a restricted range of stimuli; further it is usually rather unlikely that the underlying relationship has an 'inclined plane' form, i.e. a range with zero probability of success, a linear portion of positive slope and a range with unit probability of success; the general shape illustrated in Fig. 1.1(a) is much more common. The use of a linear model is likely to mean that even small-scale extrapolation is hazardous and that a different experimenter, using a different range of stimuli, would be likely to find an apparently different stimulus–response relationship. Similar remarks apply to the other examples of Section 1.2.

Of course, all postulated models have at best limited and approximate validity. The point here is that the use of a model, the nature of whose limitations can be foreseen, is not wise, except for very restricted purposes.

We now turn to models in which the constraint (1.4) is automatically satisfied.

We shall use the notion of a distribution of a latent response variable to motivate some alternatives. Suppose then that for a given vector x of explanatory variables the latent variable, U, has a continuous cumulative distribution function $F(u; x)$ and that the binary response $Y = 1$ is recorded if and only if $U > 0$. That is

$$\theta = \text{prob}(Y = 1; x) = 1 - F(0; x). \tag{1.5}$$

Note that since U is not directly observed there is no loss of generality in taking the critical point to be zero and that also without loss of generality we may take the standard deviation of U or some other

def u is normal, (1.5) become

$$\theta = \text{prob}(Y = 1 | x) = 1 - \underline{\Phi}(0; x)$$

Handwritten (top):

$$\theta = \text{prob}(Y=1; x) = \text{prob}(u > 0; x)$$
$$= (2\pi)^{-1/2} \int_0^\infty \exp\left[-\tfrac{1}{2}(u-\alpha-x\beta)^2\right] d u$$
$$= (2\pi)^{-1/2} \int_{-\infty}^0 \exp\left[-\tfrac{1}{2}(u-\alpha-x\beta)^2\right] d u = (2\pi)^{-1/2} \int_{-\infty}^{\alpha+x\beta} \exp\left(-\tfrac{1}{2}v^2\right)$$

Margin (top right):
change of var
$v = u - \alpha - x\beta$
$u = v + \alpha + x\beta$
dv

measure of dispersion, if constant, to be unity. If U is normal with mean $\alpha + x\beta$ it follows that

$$\theta = \Phi(\alpha + x\beta), \qquad (1.6)$$

where $\Phi(.)$ is the standard normal integral

$$\Phi(t) = (2\pi)^{-1/2} \int_{-\infty}^{t} \exp\left(-\tfrac{1}{2}v^2\right) dv. \qquad (1.7)$$

In particular if x consists of a single (scalar) component, then

$$\theta = \Phi(\alpha + x\beta) = \Phi\{(x - \mu)/\sigma\}, \qquad (1.8)$$

say, Equations (1.6) and (1.8) define the probit model which has a very extensive literature, especially in connection with bioassay (Finney, 1952). Note that the relation is linearized by the inverse normal transformation

$$\Phi^{-1}(\theta) = \alpha + x_i\beta,$$

i.e. the model can be written in a form analogous to (1.3), namely

$$\Phi^{-1}(\theta_i) = \alpha + x\beta,$$

where x_i and β are in general vectors.

We have in this derivation regarded the critical level of U as fixed and the distribution of U as changing with x. It is clear that one could have used a complementary formulation in which the distribution of U is fixed and the critical level varies with x. This second version is rather more natural in bioassay when dose, or log dose, is the explanatory variable. Here V can be taken as the dose that would just produce a response, sometimes called the tolerance, and $Y = 1$ if and only if the dose actually employed exceeds that tolerance. Note that in this formulation, if the 'dose' is $\alpha + x\beta$, then

$$\text{prob}(Y = 1; x) = \text{prob}(V \leqslant \alpha + x\beta), \qquad (1.9)$$

Handwritten (right margin):
$$= (2\pi)^{-1/2} \int_{-\infty}^{\alpha+x\beta} \exp\left(-\tfrac{1}{2}v^2\right) dv$$

relating the probability that $Y = 1$ directly to the distribution function of V. In particular if x is scalar and $\beta > 0$, then the curve of $\text{prob}(Y = 1; x)$ versus x is quite directly related to the distribution function of V.

We shall, however, for the most part use the first formulation because U thereby is more directly related to the observed binary response. An econometric illustration (Amemiya, 1985, p. 269) concerns the choice between two methods of travelling to work. Here U

might be the difference between the utilities of the two methods, assumed to have normal-theory linear regression on explanatory variables. The sign of U determines the preferred method.

The normal form is only one possibility for the distribution of U. Another is the logistic distribution with location $\alpha + x\beta$ and unit scale. This has cumulative distribution function

$$\exp(u - \alpha - x\beta)/\{1 + \exp(u - \alpha - x\beta)\}, \qquad (1.10)$$

so that

$$F(0; x) = 1/\{1 + \exp(\alpha + x\beta)\},$$

from which it follows that

$$\begin{aligned} \theta = \mathrm{prob}(Y = 1; x) = \exp(\alpha + x\beta)/\{1 + \exp(\alpha + x\beta)\}, \\ 1 - \theta = \mathrm{prob}(Y = 0; x) = 1/\{1 + \exp(\alpha + x\beta)\}. \end{aligned} \qquad (1.11)$$

Note that this relation is linearized by the transformation

$$\log\{\theta/(1 - \theta)\} = \alpha + x\beta. \qquad (1.12)$$

For scalar x and $\beta > 0$ (1.11) defines via (1.9) a probability density function on differentiation with respect to x, namely

$$\frac{\beta e^{\alpha + x\beta}}{(1 + e^{\alpha + x\beta})^2}.$$

In a reparametrized form this is

$$\frac{\exp[(x - \mu)/\tau]}{\tau\{1 + \exp[(x - \mu)/\tau]\}^2}. \qquad (1.13)$$

This is called the logistic probability density function; it is a symmetrical distribution of mean μ and standard deviation $\pi\tau/\sqrt{3}$, having a longer tail than a matching normal distribution. Statistical analysis of a random sample of continuous observations from (1.13) is not particularly simple, because (1.13) is not within the exponential family of distributions. From a manipulative point of view, the distribution has, however, the property that both density and cumulative distribution functions are expressed explicitly in terms of elementary functions. This is sometimes useful, for example in dealing with censored data.

In particular, one might have a mixture of binary data, observing success or failure at a set of preassigned x values, with a second set of data in which a sample of tolerances is measured directly. A combined analysis can be made by maximum likelihood methods, although a graphical or numerical test of the consistency of the two sets of data is desirable.

Another possibility is that response is related to the occurrence of a point in a Poisson process of rate, λ, say. If for a scalar x, we observe $Y = 1$ if and only if a point has occurred in the Poisson process before 'time' x, then

$$\text{prob}(Y = 1) = 1 - e^{-\lambda x}, \qquad (1.14)$$

with linearizing transformation

$$\log(1 - \theta) = -\lambda x.$$

This relates Y to a particular kind of extreme value problem, the occurrence of the first point. If V has the distribution of the minimum of a large number of random variables with an exponential tail, then V will have a Gumbel distribution with distribution function of the form

$$1 - \exp\{-\exp[(x - v)/\tau]\}, \qquad (1.15)$$

implying that

$$\log\{-\log(1 - \theta)\}$$

is a linear function of x. Similarly if V corresponds to the maximum of a large number of such random variables, then the distribution function is

$$\exp\{-\exp[-(x - v)/\tau]\}, \qquad (1.16)$$

so that

$$-\log(-\log\theta)$$

is a linear function of x. Now (1.15) and (1.16) are in effect related by interchanging $Y = 0$ with $Y = 1$, so that with this form of model essentially different answers are obtained by interchanging 'successes' and 'failures', in contrast with the symmetric logistic and probit models.

In most situations the concept of a latent distribution is unnecessary and it is preferable to work directly with the probabilities of success. The main exceptions occur when the latent variable has an intrinsic physical significance and when the idea is useful in suggesting models for more complex problems.

The representations discussed above are all related to linear regression via a linearizing transformation, taking the form

$$h(\theta) = \alpha + x\beta,$$

for some suitable function $h(\theta)$. In the spacial terminology of GLIM, $h(\theta)$ is called a link function. Another linearizing function that is occasionally useful via its statistical properties more than its general plausibility is the angular function

$$h(\theta) = \sin^{-1}(\sqrt{\theta}). \tag{1.17}$$

In very particular cases it may be reasonable to use non-linear functions of the components of x and unknown parameters rather than a linear function.

1.4 Some statistical considerations

In the previous section we considered various simple forms for the concise description of smooth relations between probabilities for binary response variables and explanatory variables. We now turn to a few general statistical considerations.

For the linear model (1.3) a simple possibility is to apply the method of least squares directly to the binary observations, i.e. to treat the observations 0 and 1 just as if they were quantitative observations. This is a computationally simple method which can be useful, particularly when only relatively small changes in θ_i are involved.

The method has the following limitations. Since Y_i takes only the values 0 and 1, $Y_i^2 = Y_i$ and

$$\text{var}(Y_i) = \theta_i(1 - \theta_i). \tag{1.18}$$

That is, the condition of constant variance, required even for the second-order theory of least squares, cannot hold, except in the rather uninteresting case when the θ_i's are all the same. However, it is known that quite appreciable changes in var(Y_i) induce only a modest loss of efficiency. Further, at least in the range, say, $0.2 \leqslant \theta_i \leqslant 0.8$, the function $\theta_i(1 - \theta_i)$ changes relatively little. Therefore, within this range, there is unlikely to be a serious loss of efficiency arising from the changes in var(Y_i).

The average variance of the Y_i's can be estimated approximately from the data in various ways, for example by

$$\text{ave}(Y_i)\{1 - \text{ave}(Y_i)\}, \tag{1.19}$$

where $\text{ave}(Y_i)$ is the overall proportion of 1s in the whole data. The estimate (1.19) can thus be used as an approximate theoretical residual mean square in the standard analysis of variance; a comparison with it of an empirical residual mean square provides a test of the adequacy of the model.

However, if the values of $\theta_i(1 - \theta_i)$ vary appreciably with i, there may be a serious loss of information in using the unweighted least squares estimates. One possible development is to use an iterative scheme in which fitted values $\hat{Y}_i$ are obtained from the least squares fit and then weights $\{\hat{Y}_i(1 - \hat{Y}_i)\}^{-1}$ are used in a weighted least squares analysis. This is directly related to a maximum likelihood analysis of the model; see Appendix A1.4.

A second closely related matter is that because the Y_i's are not normally distributed, no method of estimation that is linear in the Y_i's will in general be fully efficient.

It will turn out that in many ways the most useful analogue for binary response data of the linear model for normally distributed data is provided by the linear logistic model (1.11). In this for the ith individual

$$\theta_i = \frac{\exp(\alpha + x_i\beta)}{1 + \exp(\alpha + x_i\beta)}, \tag{1.20}$$

or equivalently

$$\lambda_i = \log\left(\frac{\theta_i}{1 - \theta_i}\right) = \alpha + x_i\beta. \tag{1.21}$$

From a rather theoretical point of view the close connection of this with the exponential family brings about considerable advantages; see Appendix A1.4.

In fact the linear logistic expression provides a flexible collection of models that are at least potentially capable of representing a range of situations involving binary responses. Broadly, the linear logistic models, and the questions they are intended to answer, are analogous to the normal-theory linear models, which underlie the techniques of analysis of variance and covariance, and simple and multiple regression. Of course, not all problems about approximately normally distributed observations are best tackled via a linear model; correspondingly it is quite wrong to think that a linear logistic model will be successful every time binary responses are encountered.

1.5 Some numerical comparisons

To understand the practical meaning of an analysis based on the logistic model, it is important to appreciate the meaning of differences on a logistic scale. To help in this, the following properties are useful.

The second column of Table 1.9 gives the standardized logistic function

$$x = \log\left(\frac{\Lambda(x)}{1-\Lambda(x)}\right) \Longleftrightarrow \qquad \Lambda(x) = e^x/(1 + e^x). \tag{1.22}$$

This is helpful in interpreting the parameters of the logistic response curve for a scalar explanatory variable, x. Since $\Lambda(1) = 0.731$, the parameter β in the form

$$\frac{e^{\alpha + x\beta}}{1 + e^{\alpha + x\beta}}$$

is such that $1/\beta$ is approximately the distance in x units between the 75% point and the 50% point of the curve. Similarly the distance between the 95% point and the 50% point is approximately $3/\beta$, etc.

The difference $\Delta\theta$ of two probabilities corresponding to a difference $\Delta\lambda$ on a logistic scale depends on the particular values of the probabilities involved. Approximately, however, we have on differentiating (1.22) that

$$\Delta\lambda \sim \frac{\Delta\theta}{\theta(1-\theta)}. \tag{1.23}$$

Thus near $\theta = \frac{1}{2}$, $\Delta\lambda \sim 4\Delta\theta$, whereas for small θ, $\Delta\lambda \sim \Delta\theta/\theta$, and for θ very near 1, $\Delta\lambda \sim \Delta\theta/(1 - \theta)$. Alternatively, differences on a logistic scale can be interpreted easily by thinking from the beginning not in terms of the probability θ of success but rather in terms of the odds for success against failure, namely $\theta/(1 - \theta)$. The exponential of a logistic difference is thus a ratio of odds. If $\Delta\lambda = 1$, the odds in the second situation are about 2.7 times those in the first situation.

To compare the mathematical form of different transformations we can, without loss of generality, suppose that the 50% point is at $x = 0$

and we therefore write the logistic curve as

$$\frac{e^{\beta x}}{1 + e^{\beta x}}. \qquad (1.24)$$

For comparison, suppose that the probability of success varies linearly with x according to the function

$$\begin{cases} 1 & (x > \tfrac{1}{2}\gamma^{-1}), \\ \tfrac{1}{2} + \gamma x & (|x| \leqslant \tfrac{1}{2}\gamma^{-1}), \\ 0 & (x < -\tfrac{1}{2}\gamma^{-1}). \end{cases} \qquad (1.25)$$

Two further relationships correspond to the integrated normal and the angular transforms. According to the first, the probability of success at stimulus x is

$$\Phi(\xi x). \qquad (1.26)$$

When the angular transform is used, we take the probability of success at stimulus x to be

$$\begin{cases} 1 & (x > \tfrac{1}{4}\pi/\eta), \\ \sin^2(\eta x + \tfrac{1}{4}\pi) & (|x| \leqslant \tfrac{1}{4}\pi/\eta), \\ 0 & (x < -\tfrac{1}{4}\pi/\eta). \end{cases} \qquad (1.27)$$

In (1.24)–(1.27), the adjustable parameters β, γ, ξ and η allow the representation of relationships of different slopes.

If $\theta(x)$ denotes the probability of success at level x, the linearizing transformations associated with the response curves (1.24)–(1.27) are respectively from $\theta(x)$ to

$$\log\left\{\frac{\theta(x)}{1 - \theta(x)}\right\}, \quad \theta(x), \quad \Phi^{-1}\{\theta(x)\}, \quad \sin^{-1}\{\sqrt{\theta(x)}\}. \quad (1.28)$$

To compare the four response curves, all of which are symmetric, we have to choose the scale constants β, γ, ξ and η in comparable form. This has been done rather arbitrarily by taking $\beta = 1$ and then choosing γ, ξ and η so that all four curves agree at the 80% point. We get $\gamma = 0.144$, $\xi = 0.607$ and $\eta = 0.232$. Table 1.9 gives the resulting functions; they all behave symmetrically about $x = 0$.

For most purposes the logistic and the normal agree closely over the whole range. The only exception is when special interest attaches to the regions where the probability of success is either very small, or

Table 1.9 *Comparison of probability of success as given by four stimulus response curves*

x	Logistic	Normal	Angular	Linear
0	0.500	0.500	0.500	0.500
0.5	0.622	0.619	0.615	0.608
1.0	0.731	0.728	0.724	0.716
1.5	0.818	0.818	0.821	0.825
2.0	0.881	0.887	0.900	0.933
2.5	0.924	0.935	0.958	1.000
3.0	0.953	0.965	0.992	1.000
3.5	0.971	0.983	1.000	1.000
4.0	0.982	0.992	1.000	1.000
4.5	0.989	0.997	1.000	1.000
5.0	0.993	0.999	1.000	1.000

very near one. Then the normal curve approaches its limit more rapidly than the logistic. The angular and linear relations agree with the other two reasonably well when the probability of success is in the range 0.1–0.9. Outside this range the finite limits on the last two curves usually seriously restrict their usefulness. Subject to these provisos, an analysis in terms of any of the four relations is likely to give virtually equivalent results (Claringbold, Biggers and Emmens, 1953; Naylor, 1964).

We have considered the general use of a linear model in the probabilities in Section 1.4. The statistical usefulness of the angular transformation arises from its variance-stabilizing property. With a number of groups of observations of equal size, fitting by unweighted least squares is appropriate. Both linear and angular functions are, however, severely limited for general usefulness by their finite range. The use of the integrated normal curve in connection with binary data has been extensively discussed (Finney, 1952). The absence of sufficient statistics is a theoretical disadvantage, primarily with relatively simple problems, where a simple solution is often available for the logistic and not for the integrated normal.

The above discussion is in terms of a stimulus–response curve. The general formulation would be in terms of linear models in which one or other of the transforms (1.28) is specified by a linear expression in unknown parameters.

In particular the close numerical equivalence between logistic and probit functions implies that parameters estimated from one model can to a close approximation be converted into equivalent values for the other model.

1.6 Some generalizations of the logistic model

It is possible to generalize all the above forms in various ways and we now outline one or two possibilities. Aranda-Ordaz (1981) proposed two families of linearizing transformations which can easily be inverted and which span a range of forms. The first, which is restricted to symmetric cases, i.e. those which are essentially invariant to interchanging successes and failures, is

$$h^{(s)}(\theta) = \frac{2\,\theta^v - (1-\theta)^v}{v\,\theta^v + (1-\theta)^v}. \tag{1.29}$$

This gives the logistic as the limiting case $v = 0$, the linear as $v = 1$; numerical work shows that $v = 0.39$ is very close to the probit and $v = 0.67$ very close to the angular transformation. The second family has

$$h^{(a)}(\theta) = \log\left[\{(1-\theta)^{-v} - 1\}/v\right]. \tag{1.30}$$

This includes the extreme value model ($v = 0$) and the logistic ($v = 1$) as special cases. Where there is some doubt about the appropriate transformation a rather formal approach is to use one or both of the above as the linearizing transformation and to fit the resulting model by maximum likelihood for a series of values of v, plotting the maximized log likelihood against v, i.e. obtaining the profile log likelihood of v.

The models discussed above all have the property that for extreme values of the explanatory variables the limiting probability that $Y = 1$ is either 0 or 1. If this is not the case there are two broad possibilities. One noted above is to replace the linear combination of explanatory variables by a non-linear combination, for example $\alpha + \gamma e^{-x\delta}$, retaining the linearizing function. The other is to modify the linearizing transformation. One explicit motivation is by some notion of misclassification. For example suppose that an unobserved binary response variable Γ is governed by a linear logistic model (1.12). Suppose further that with probability π_{10} the response $\Gamma = 1$ is recorded as $Y = 0$ and that with probability π_{01} the response $\Gamma = 0$ is recorded as $Y = 1$. Then

$$\text{prob}(Y = 1; x) = \{\pi_{01} + (1 - \pi_{10})\exp(\alpha + x\beta)\}/\{1 + \exp(\alpha + x\beta)\}.$$
$$(1.31)$$

This relation has non-zero and non-unit asymptotes as $x\beta \to \pm \infty$. Obvious special cases have one of the π's zero. Note that (1.31) has linearizing transformation

$$\log\{(\theta - \pi_{01})/(1 - \pi_{10} - \theta)\}.$$

If the ε's are regarded as known this raises no special problems, but in applications quite often the ε's have to be estimated. Note finally that we have given just one motivation of (1.31); the same mathematical form can be used whenever non-zero asymptotes are encountered. Thus Mr D. Sampson of Imperial College in as yet unpublished work analysed a large number of sets of data involving the relation between the length of female whales and the proportion pregnant. Fitting of (1.31) with $\varepsilon_0 = 0$ gave a convenient way of summarizing each set of data via a small number of interpretable parameters.

Bibliographic notes

The earliest work on binary and more generally qualitative data centred on two themes, the chi-squared test of null hypotheses of no association or no dependence (Pearson, 1900), and the construction of indices of association; for a thorough review of the latter, see Goodman and Kruskal (1954, 1959, 1963). The introduction of specific models for dependence is much more recent, and in a sense stems from the analysis of dose–response curves in bioassay (Finney, 1952, 1964) and more specifically from the use of maximum likelihood to estimate the 50% point of such curves (Fisher, 1935).

The importance of the logistic distribution in bioassay and the advantages of estimates obtained by weighted least squares applied to transformed values (minimum logit chi-squared) were stressed in a series of papers by Berkson (1944, 1951, 1953, 1955a, b, 1957, 1960, 1968). Another use of the empirical logit transform was treated by Woolf (1955).

Much of the early literature on logistic binary regression focused largely on the logistic curve as an alternative to the probit (integrated normal), the discussion of relative merits being somewhat confounded with a dispute about maximum likelihood versus empirically

weighted least squares (minimum logit chi-squared) as point estimation procedures for slope and 50% point. Modifications to the
empirical logit transform to remove bias originate with Haldane
(1955) and have been very thoroughly studied by Gart, Pettigrew and
Thomas (1985). The systematic use of empirically weighted least
squares to analyse various linear logistic regression problems has
been studied by Bhapkar (1961, 1968), Grizzle (1961) and Bhapkar
and Koch (1968). For the design of experiments to discriminate
between alternative models, e.g. logistic and probit, see Chambers
and Cox (1967).

The broader use of logistic regression probably dates from the
work connected with the Framingham study (a major US hyperten-
sion investigation), especially the contributions of J. Cornfield. Cox
(1958a) related the model to the exponential family of distributions
and derived, in a unified way, a variety of tests and interval estimates.

For further discussion of transformations embracing logistic and
other relations, see Guerrero and Johnson (1982) and Stukel (1988).

CHAPTER 2

Special logistic analyses

2.1 Introduction

2.1.1 Formulation

The main object of the present chapter is to show how the linear logistic model (1.11) can be a basis for the systematic discussion of a range of problems connected with binary response variables. The special cases are intended both to be of intrinsic importance and also to serve as illustrations of formulation and hence as a guide to the discussion of non-standard situations.

The starting point is a set of binary responses represented by random variables $Y_1, \ldots, Y_n$ with $\theta_i = \text{prob}(Y_i = 1)$. For each $i = 1, \ldots, n$ there is a row vector $x_i = (x_{i1}, \ldots, x_{ip})$ of explanatory variables. In the linear logistic model there are unknown parameters α and the $p \times 1$ column vector β such that

$$\theta_i = \frac{e^{\alpha + x_i\beta}}{1 + e^{\alpha + x_i\beta}}, \quad 1 - \theta_i = \frac{1}{1 + e^{\alpha + x_i\beta}},$$

$$\lambda_i = \log\left(\frac{\theta_i}{1 - \theta_i}\right) = \alpha + x_i\beta = \alpha + \sum_{t=1}^{p} x_{it}\beta_t. \tag{2.1}$$

Except in the discussion of time series we shall assume that $Y_1, \ldots, Y_n$, referring to n distinct individuals, are mutually independent.

For some general purposes it is convenient to change the notation slightly by writing $\beta_0 = \alpha$, $x_{i0} = 1$, when (2.1) is equivalent to

$$\lambda = x\beta, \tag{2.2}$$

where λ is an $n \times 1$ column, x is an $n \times d$ matrix, $d = p + 1$, and β is a $d \times 1$ column of parameters, $\beta^{\text{T}} = (\beta_0, \beta_1, \ldots, \beta_p)$. It will be clear from the context which form is being used. The reader unfamiliar with these representations should concentrate first on the simple special cases to be discussed in Sections 2.2 and 2.3.

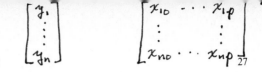

$$\begin{bmatrix} y_1 \\ \vdots \\ y_n \end{bmatrix} \qquad \begin{bmatrix} x_{10} & \cdots & x_{1p} \\ \vdots & & \vdots \\ x_{n0} & \cdots & x_{np} \end{bmatrix}$$

We shall use three broad approaches for the development of methods of analysis, namely (a) the calculation of 'exact' conditional distributions, (b) the use of maximum likelihood, (c) the calculation of empirical logistic transforms. These are put in a more general setting in Appendix A1.3 and A1.4.

The key to (a) and (b) is the likelihood function.

2.1.2 Likelihood function

Because of the assumed independence of the responses on different individuals the likelihood, based on (2.1), of an observed binary sequence $y_1, \ldots, y_n$ is $\left(e^{\alpha + x_1\beta}\right)\cdots\left(e^{\alpha + x_n\beta}\right)\Big/ \prod\left(1 + e^{\alpha + x_i\beta}\right)$

$$\mathrm{prob}(Y_1 = y_1, \ldots, Y_n = y_n) = \frac{\exp\left(\sum_{s=0}^{p}\beta_s t_s\right)}{\prod_{i=1}^{n}(1 + e^{x_i\beta})}, \tag{2.3}$$

where

$$t_s = \sum_{i=1}^{n} x_{is} y_i \tag{2.4}$$

is the observed value of the random variable $T_s = \sum x_{is} Y_i$. The sufficient statistics $t_0, \ldots, t_p$ are sub-totals of the columns of the matrix formed from the n rows x_i, the elements summed corresponding to the rows in which successes occur.

Note that because $x_{i0} = 1$, $T_0 = \sum y_i$ is the total number of successes. The distribution of $T_0, \ldots, T_p$ follows on summing (2.4) (2.3) over all binary sequences that generate the particular values $t_0, \ldots, t_p$. In fact

$$\mathrm{prob}(T_0 = t_0, \ldots, T_p = t_p) = \frac{c(t_0, \ldots, t_p)\exp\left(\sum_{s=0}^{p}\beta_s t_s\right)}{\prod_{i=1}^{n}(1 + e^{x_i\beta})} \tag{2.5}$$

where $c(t_0, \ldots, t_p)$ is the number of distinct binary sequences that yield the specified values $t_0, \ldots, t_p$. A generating function for $c(t_0, \ldots, t_p)$ is

$$\begin{aligned} C(\zeta_0, \ldots, \zeta_p) &= \sum c(t_0, \ldots, t_p)\zeta_0^{t_0} \cdots \zeta_p^{t_p} \\ &= \prod_{i=1}^{n}\left(1 + \zeta_0^{x_{i0}} \cdots \zeta_p^{x_{ip}}\right). \end{aligned} \tag{2.6}$$

2.1.3 Conditional inference

Now suppose that we are interested in one of the regression parameters, regarding the remainder as nuisance parameters. Without loss of generality we suppose the parameter of interest to be β_p, the nuisance parameters being $\beta_0, \ldots, \beta_{p-1}$. To summarize what the data tell us about β_p we consider the conditional distribution of T_p given the observed values of $T_0, \ldots, T_{p-1}$. There are two lines of argument that lead to this. In both approaches we need consider only methods involving the sufficient statistics. First, we may look for a distribution that depends on the value of β_p, but not on the nuisance parameters $\beta_0, \ldots, \beta_{p-1}$. For any fixed and known value of β_p, a sufficient statistic for the remaining parameters is $(T_0, \ldots, T_{p-1})$. Therefore the distribution of the observations, and hence also of T_p, given that $T_0 = t_0, \ldots, T_{p-1} = t_{p-1}$, cannot involve $\beta_0, \ldots, \beta_{p-1}$. Further it can be shown (Lehmann, 1986, p. 145) that the use of this conditional distribution is the only way to obtain exact independence of the nuisance parameters. This is the standard Neyman–Pearson approach to the elimination of nuisance parameters. It leaves open the possibility that a more sensitive analysis of the data may be available by adopting a less drastic requirement than that the probability properties should be completely independent of the nuisance parameters.

The second approach to the conditional distribution (Fisher, 1956, Section IV.4) stems from an idea which is conceptually very appealing in special cases, but difficult to formalize in general. This is that if we were given only the values $t_0, \ldots, t_{p-1}$, no conclusions could be drawn about β_p. The values $(t_0, \ldots, t_{p-1})$ determine the precision with which conclusions about β_p can be drawn and it is thus appropriate to argue conditionally on the observed values. This is to ensure that we attach to the conclusions the precision actually achieved and not that to be achieved hypothetically in a recognizably distinct situation that has in fact not occurred.

Fortunately, the two lines of argument point to the same conclusion in the present situation.

2.1.4 Properties of conditional distributions

To find the conditional distribution of T_p given $T_0 = t_0, \ldots, T_{p-1} = t_{p-1}$, we have that

$$\text{prob}(T_p = t_p \mid T_0 = t_0, \ldots, T_{p-1} = t_{p-1})$$
$$= \frac{\text{prob}(T_0 = t_0, \ldots, T_p = t_p)}{\text{prob}(T_0 = t_0, \ldots, T_{p-1} = t_{p-1})}. \tag{2.7}$$

The numerator is given by (2.5) and the denominator by summing (2.5) over all possible t_p. Hence, when we form the ratio (2.7), several factors cancel and the conditional probability is

$$\frac{c(t_0, \ldots, t_p)e^{\beta_p t_p}}{\sum_u c(t_0, \ldots, t_{p-1}, u)e^{\beta_p u}}. \tag{2.8}$$

Note that (2.8) does not involve $\beta_1, \ldots, \beta_{p-1}$; indeed this fact is a consequence of the sufficiency of the conditioning statistics.

Most of the subsequent discussion centres on (2.8). It is convenient to simplify the notation. We denote the parameter β_p of interest by β, the statistic t_p by t and the conditioning statistic $(t_0, \ldots, t_{p-1})$ by $t_{(p-1)}$. The distribution (2.8) can then be written

$$p_T(t; \beta) = \frac{c(t_{(p-1)}, t)e^{\beta t}}{\sum_u c(t_{(p-1)}, u)e^{\beta u}}. \tag{2.9}$$

An important special case of (2.9) corresponds to $\beta = 0$,

$$p_T(t; 0) = \frac{c(t_{(p-1)}, t)}{\sum_u c(t_{(p-1)}, u)}, \tag{2.10}$$

so that the distribution is determined by the combinatorial coefficients.

We denote the moment generating function of (2.8) by

$$M_T(z; \beta) = \sum_t e^{zt} p_T(t; \beta) = \frac{\sum_t c(t_{(p-1)}, t)e^{(\beta + z)t}}{\sum_u c(t_{(p-1)}, u)e^{\beta u}}.$$

In particular, when $\beta = 0$,

$$M_T(z; 0) = \frac{\sum_t c(t_{(p-1)}, t)e^{zt}}{\sum_u c(t_{(p-1)}, u)}. \tag{2.11}$$

Thus

$$M_T(z; \beta) = \frac{M_T(z + \beta; 0)}{M_T(\beta; 0)}, \tag{2.12}$$

or, in terms of cumulant generating functions,

$$K_T(z; \beta) = K_T(z + \beta; 0) - K_T(\beta; 0), \tag{2.13}$$

where $K_T(z; \beta) = \log M_T(z; \beta)$. The potential usefulness of (2.12) and (2.13) lies in the possibility of obtaining properties of the distribution of T for non-zero β from properties for $\beta = 0$.

2.1.5 Optimum properties

To complete the formal theoretical analysis, consider the problem of testing the null hypothesis $\beta = \beta_0$ against an alternative $\beta = \beta'$, with, say, $\beta' > \beta_0$. We confine attention to the sufficient statistics and, for the reasons outlined in Section 2.1.3, consider the conditional distribution (2.9). To obtain a most powerful test we therefore apply the Neyman–Pearson lemma to (2.9), i.e. form a critical region from those sample points having large values of the likelihood ratio

$$\frac{p_T(t; \beta')}{p_T(t; \beta_0)} \propto e^{(\beta' - \beta_0)t}, \tag{2.14}$$

the factor of proportionality being independent of t. Thus, for all $\beta' > \beta_0$, the critical region should consist of the upper tail of values of t and the resulting procedure is uniformly most powerful.

Corresponding to an observed value t, the one-sided significance level P_+ against alternatives $\beta > \beta_0$ is thus

$$P_+(t; \beta_0) = \sum_{u \geq t} p_T(u; \beta_0) \tag{2.15}$$

and against alternatives $\beta < \beta_0$ the level is, correspondingly,

$$P_-(t; \beta_0) = \sum_{u \leq t} p_T(u; \beta_0). \tag{2.16}$$

For two-sided alternatives it is usual to quote the value

$$P(t; \beta_0) = 2 \min \{P_+(t; \beta_0), P_-(t; \beta_0)\}. \tag{2.17}$$

To obtain an upper $(1 - \varepsilon)$ confidence limit for β, corresponding to an observed value t, we take the largest value of β, say $\beta_+(t)$, which is just significantly too large at the ε level, i.e. the solution of

$$\varepsilon = P_-\{t; \beta_+(t)\}. \tag{2.18}$$

Similarly a lower confidence limit is obtained from

$$\varepsilon = P_+\{t; \beta_-(t)\}. \tag{2.19}$$

Thus, for given ε, only a discrete set of values is possible for the confidence limits, because the possible values of t are discrete. This can be shown to imply that if the true value of β equals one of the attainable limits, then the probability of error is the nominal value ε, and that otherwise the probability of error is less than ε. The confidence coefficient for the interval $(\beta_-(t), \beta_+(t))$ is thus at least

$(1 - 2\varepsilon)$. Note that if we are interested in the consistency of the data with a particular value of β, say β_0, we are concerned with (2.15)–(2.17), rather than with trying to achieve an arbitrary pre-assigned value of ε.

2.1.6 Empirical logistic transform

In the general formulation the values of the explanatory variables for different individuals, i.e. different values of i, are allowed to be distinct. Sometimes, however, the individuals can be arranged in groups such that all individuals within a group have the same values of all explanatory variables and hence, under the models considered here, have the same probability of success.

Provided that the responses of different individuals are mutually independent, it follows from the above, and is obvious on general grounds, that for any group the only relevant aspects of the response are the number of individuals, m, say, and the total number of successes, represented by the random variable R. If ϕ is the probability of success for the group, then R has a binomial distribution of index m and probability ϕ, so that in particular

$$E(R) = m\phi, \qquad \text{var}(R) = m\phi(1 - \phi),$$
$$E(R/m) = \phi, \qquad \text{var}(R/m) = \phi(1 - \phi)/m. \qquad (2.20)$$

Now in the present context, interest centres not so much on ϕ as on the logistic transform

$$\lambda = \log\{\phi/(1 - \phi)\}.$$

Provided that neither the number of successes nor the number of failures is too small, λ is reasonably estimated by

$$Z = \log\{R/(m - R)\} = \log\left\{\frac{R/m}{1 - R/m}\right\} \qquad (2.21)$$

which we call the empirical logit transform.

More generally if the parametric function of interest is $h(\phi)$, then we consider $h(R/m)$. Now provided that the variation in R/m is relatively small we can write

$$h(R/m) \simeq h(\phi) + (R/m - \phi)h'(\phi) \qquad (2.22)$$

from which it can be shown that $h(R/m)$ is approximately normally distributed with mean $h(\phi)$ and variance

$$\{h'(\phi)\}^2 \text{var}(R/m) = \{h'(\phi)\}^2 \phi(1 - \phi)/m. \qquad (2.23)$$

$$\text{def } h(\phi) = \log\left(\frac{\phi}{1-\phi}\right)$$

$$\text{Then } h'(\phi) = \frac{1-\phi}{\phi}\left[+(1-\phi)^{-1} + \phi(1-\phi)^{-2}\right]$$

$$= + \frac{1}{\phi(1-\phi)}$$

$$Z = \log\left\{\frac{R}{m-R}\right\} \quad \text{``empirical logit transform''}$$

In particular Z is approximately normally distributed with mean λ and variance $\{m\phi(1-\phi)\}^{-1}$. This variance is itself estimated by replacing ϕ by R/m to give

$$V = m/\{R(m-R)\}. \tag{2.24}$$

If we take the expansion (2.22) to a further term by writing

$$h(R/m) \simeq h(\phi) + (R/m - \phi)h'(\phi) + \tfrac{1}{2}(R/m - \phi)^2 h''(\phi),$$

we have

$$E\{h(R/m)\} \simeq h(\phi) + \tfrac{1}{2}\phi(1-\phi)h''(\phi)/m. \tag{2.25}$$

The bias implied by the second term can be corrected most easily in the case of the empirical logit transform by replacing Z by

$$Z(a) = \log\{(R+a)/(m-R+a)\}$$

for some constant a. A calculation analogous to (2.25) shows that

$$E\{Z(a)\} \simeq \lambda + \frac{(1-2\phi)(a-\tfrac{1}{2})}{m\phi(1-\phi)}$$

leading to the choice $a = \tfrac{1}{2}$, i.e. to a modification of the definition (2.21) to

$$Z = \log\{(R+\tfrac{1}{2})/(m-R+\tfrac{1}{2})\}. \tag{2.26}$$

This has the further advantage that a singularity in the definition at $R = 0$ or m is avoided. The variance V can be modified similarly to

$$V = \frac{(m+1)(m+2)}{m(R+1)(m-R+1)}. \tag{2.27}$$

There will thus be a pair (Z, V) for each group of individuals.

The above modification (2.26) of the empirical logistic transform and the associated variance (2.27) are appropriate if unweighted linear combinations of the transforms are to be used. It can be shown that if a weighted least squares analysis of the transforms is to be used it is preferable to take

$$Z^{(w)} = \log\left(\frac{R-\tfrac{1}{2}}{m-R-\tfrac{1}{2}}\right), \qquad V^{(w)} = \frac{(m-1)}{R(m-R)}; \tag{2.28}$$

note that this leads to groups with $R = 0$ or $R = m$ making no contribution.

Following Gart and Zweifel (1967), McCullagh and Nelder (1989, p.107) recommend

$$V = \frac{(m+1)}{(R+\tfrac{1}{2})(m-R+\tfrac{1}{2})}$$

In the first edition of this book some prominence was given to methods of analysis using empirical logit transforms, usually via a form of weighted least squares. This had the advantage of avoiding iterative calculations, but, with the wide availability of computer packages for iterative fitting, we have in the present account laid less emphasis on empirical transforms, although they remain important both for quick calculations and also for non-standard problems.

2.2 Simple regression

2.2.1 Preliminaries

Suppose that to each individual is attached a single explanatory variable x thought to influence the probability of success. The data are represented by n pairs $(x_1, Y_1), \ldots, (x_n, Y_n)$, where the Y_i's are assumed to be independent binary random variables. Examples 1.4–1.6 illustrate this situation. There are two particular cases of special interest. The first has $x_i = i$ covering, in particular, the dependence of probability of success on the serial order of the trial (Example 1.4). In the second the x_i's are grouped into a fairly small number, k, of sets, each with a constant value of the explanatory variable. This can be regarded as a $k \times 2$ contingency table (Example 1.6) in which a 'score' is attached to each row.

As in other regression problems graphical methods are very important. We can, for example, group the x values, if they are not grouped to begin with, and plot the empirical logistic transform, (2.26), against x.

2.2.2 Formal analysis

Suppose that we analyse the data in the light of the linear logistic model,

$$\lambda = \alpha \begin{pmatrix} 1 \\ \vdots \\ 1 \end{pmatrix} + \beta \begin{pmatrix} x_1 \\ \vdots \\ x_n \end{pmatrix};$$

in the problems we shall consider here, β is the parameter of primary interest. Note, however, that in the application to stimulus–response curves the model may hold but interest will not normally be primarily in β; it may, for example, be in the stimulus $-\alpha/\beta$ at which the probability of success is $\frac{1}{2}$.

The general results of Section 2.1 take on a particularly simple form here. The conditioning statistic is $\sum Y_i$, the total number of successes, its observed value being, say, t_0. The statistic associated with the parameter of interest is $\sum x_i Y_i$, i.e. the sum of the x_i's over those individuals giving a success. Given t_0, the statistic $T = \sum x_i Y_i$ is the total of a sample of size t_0 drawn from the finite population $\mathcal{X} = \{x_1, \ldots, x_n\}$. Further, in the special case when $\beta = 0$, so that the probability of success is constant, it is clear that all distinct samples of size t_0 have equal probability, i.e. that T is the total of a sample drawn randomly without replacement from the finite population $\mathcal{X}$.

Now the cumulants and moments of the null distribution of T are known; see for example, Kendall and Stuart (1963, pp. 300–4). In fact, the first four moments are

$$E(T;0) = t_0 m_1, \qquad\qquad \beta = \phi$$

$$\mathrm{var}\,(T;0) = \frac{t_0(n - t_0)m_2}{(n - 1)},$$

$$\mu_3(T;0) = \frac{t_0(n - t_0)(n - 2t_0)m_3}{(n - 1)(n - 2)}, \qquad\qquad (2.29)$$

$$\mu_4(T;0) = \frac{t_0(n - t_0)}{(n - 1)(n - 2)(n - 3)}\{(n^2 - 6nt_0 + n + 6t_0^2)m_4$$
$$+ 3(t_0 - 1)n(n - t_0 - 1)m_2^2\},$$

where

$$m_1 = \sum x_i/n, \quad m_s = \sum (x_i - m_1)^s/n \qquad (s = 2, 3, \ldots).$$

Here the zeros in $E(T;0), \ldots$, indicate that $\beta = 0$. The dependence on the condition $T_0 = t_0$ is not shown on the left-hand side.

A normal approximation to the null distribution of T will often be adequate, but it can be checked by computing the skewness and kurtosis of T and if necessary, corrections for non-normality can be introduced. The use of a normal approximation is supported by limit theorems which give under weak conditions the asymptotic normality of T, as t_0 and n tend to infinity with t_0/n fixed.

To get some insight into the rate of approach to normality, it is useful to write $f = t_0/n$ for the effective sampling fraction and to take n

large with f fixed. Then

$$E(T;0) = t_0 m_1,$$
$$\text{var}(T;0) = t_0 m_2 (1-f)\{1 + n^{-1} + o(n^{-1})\}, \quad (2.30)$$
$$\rho_3(T;0) \sim \frac{(1-2f)\rho_3}{\sqrt{\{nf(1-f)\}}},$$
$$\rho_4(T;0) \sim -\frac{6}{n} + \frac{\{1-6f(1-f)\}\rho_4}{nf(1-f)},$$

where $\rho_3 = m_3/m_2^{3/2}$ and $\rho_4 = m_4/m_2^2 - 3$ refer to the finite population, $\mathscr{X}$.

2.2.3 Special cases

There are two special cases of particular interest, corresponding respectively to tests for serial order effects (Example 1.4) and to the analysis of a $k \times 2$ contingency table for trend (Example 1.6).

In the first of these problems the finite population $\mathscr{X}$ is $\{1, 2, \ldots, n\}$. The test statistic is, under the null hypothesis, the total of a random sample of size t_0 drawn without replacement from $\mathscr{X}$. This leads to the same null probability distribution as arises in the two-sample Wilcoxon test (Kendall and Stuart, 1967, p. 492), and tables of its exact distribution (Fix and Hodges, 1955), can be used if t_0 or $n - t_0$ is small. Note that in the Wilcoxon test the 0s and 1s correspond to a coding of the two samples and the values $1, 2, \ldots, n$ are the rank numbers of the pooled data; the roles of the random and non-random variables in that problem are thus complementary to those in the regression problem.

A simple, if artificial, numerical example illustrates some of the above formulae.

Example 2.1 A series of seven individuals
The moments of the finite population $\{1, 2, \ldots, 7\}$ are

$$m_1 = \tfrac{1}{7}\sum_{i=1}^{7} i = 4, \quad m_2 = \tfrac{1}{7}\sum_{i=1}^{7} (i-4)^2 = 4, \quad m_3 = 0, \quad m_4 = 28.$$

If there are three successes in the seven trials, we are concerned with the sum of a sample of size $t_0 = 3$ from this finite population. Hence,

Table 2.1 *Comparison of exact and approximate tail areas*

Observed value t	prob$(T \geqslant t; \beta = 0)$		
	Exact	Normal approximation	Edgeworth approximation
18	$1/35 = 0.029$	0.026	0.024
17	$2/35 = 0.057$	0.056	0.057
16	$4/35 = 0.114$	0.108	0.116
15	$7/35 = 0.200$	0.188	0.200
14	$11/35 = 0.314$	0.298	0.309
13	$15/35 = 0.429$	0.430	0.435

from (2.30),

$$E(T;0) = 12, \qquad \text{var}(T;0) = 8,$$
$$\mu_3(T;0) = 0, \qquad \mu_4(T;0) = 156.8,$$

so that

$$\rho_3(T;0) = 0, \qquad \rho_4(T;0) = -11/20.$$

Table 2.1 gives the upper tail of the null distribution of T obtained by enumeration of the 35 distinct samples of size three from the finite population, or by use of a normal approximation with continuity correction, or by use of an adjustment based on the Edgeworth expansion. The effect of the correction for kurtosis is small. An alternative way of allowing for non-normality is by Table 42 of Pearson and Hartley (1966), based on the fitting of Pearson curves, but the example shows that such corrections are likely often to be unnecessary.

2.2.4 Maximum likelihood analysis

The most commonly used method of analysis for such data involves a maximum likelihood approach; see Appendix A1.5 for a review of the main ideas. This is especially appropriate when, as is usually the case, estimation as well as significance testing is called for. While confidence intervals for β can be obtained via the 'exact' approach of Section 2.2.3, the calculations are rather cumbersome.

Example 2.2 *Analysis of ingot data*

To illustrate the arguments involved, it is convenient to use p.26: $\lambda_i = \log\left(\dfrac{\theta_i}{1-\theta_i}\right) \Leftrightarrow \theta_i = \dfrac{e^{\alpha+\beta x_{1i}}}{1+e^{\alpha+\beta x_{1i}}}$

Example 1.8, concentrating on the regression on x_1, i.e. assuming that x_2 has no effect on the probability of success. Under the logistic model

$$\lambda_i = \alpha + \beta x_{1i},$$

where x_{1i} is the value of x_1 for the ith ingot, the log likelihood is

$$12\alpha + 370\beta - \{55\log(1 + e^{\alpha+7\beta}) + 157\log(1 + e^{\alpha+14\beta})$$
$$+ 159\log(1 + e^{\alpha+27\beta}) + 16\log(1 + e^{\alpha+51\beta})\}. \quad (2.31)$$

NB: $370 = 6\times7 + 2\times14 + 7\times27 + 3\times51$

This is maximized at

$$\hat{\alpha} = -5.415 \qquad \hat{\beta} = 0.0807 \qquad (2.32)$$

and the inverse of the matrix of second derivatives at the maximum likelihood point is

$$\begin{pmatrix} 0.5293 & -0.0148 \\ -0.0148 & 0.000500 \end{pmatrix}. \qquad (2.33)$$

In Section A1.5 two procedures are suggested for obtaining confidence limits. The first and simpler is to use an asymptotic standard error obtained from (2.33). That for $\hat{\beta}$ is $\sqrt{0.000500} = 0.0224$ and approximate 95% and 98% limits are thus

$$(0.037, 0.125) \qquad \text{and} \qquad (0.029, 0.133). \qquad (2.34)$$

To apply the second method further calculations are required. The likelihood (2.31) is maximized with respect to α for a range of values of β to give the profile log likelihood function

$$l(\hat{\alpha}_\beta, \beta)$$

shown in Table 2.2, where $\hat{\alpha}_\beta$ is the maximum likelihood estimate of α for given β.

Table 2.2 *Ingot data. Log likelihood maximized with respect to α for given β, i.e. profile log likelihood of β*

β	$l(\hat{\alpha}_\beta, \beta)$	β	$l(\hat{\alpha}_\beta, \beta)$
0.03	−50.12	0.09	−47.77
0.04	−49.28	0.10	−48.06
0.05	−48.61	0.11	−48.55
0.06	−48.11	0.12	−49.22
0.07	−47.80	0.13	−50.08
0.08	−47.69		

At $\beta = \hat{\beta} = 0.0807$, $l(\hat{\alpha}, \hat{\beta}) = -47.69$.

See p. 11

p.27 ⟹ likelihood

A confidence region for β is formed, according to (A1.50), from those values of β giving $l(\hat{\alpha}_\beta, \beta)$ sufficiently close to the overall maximum. For 95% and 98% regions the allowable differences from the maximum are, from the chi-squared tables with one degree of freedom,

$$\tfrac{1}{2} \times 3.841 = 1.920 \quad \text{and} \quad \tfrac{1}{2} \times 5.412 = 2.706.$$

Thus the confidence regions are those giving $l(\hat{\alpha}_\beta, \beta) \geqslant 49.61$ and 50.40; a graph plotted from Table 2.2 gives the intervals

$$(0.036, 0.124) \quad \text{and} \quad (0.027, 0.133). \tag{2.35}$$

As will usually happen, (2.34) and (2.35) are in close agreement. The general arguments for preferring the second method are outlined in Appendix A1.5.

2.2.5 Combination of regressions

A closely analogous discussion holds when there are k independent sets of data within each of which regression on an explanatory variable x is to be considered. Suppose that the logistic regression equation has the form

$$\alpha_s + \beta x_{is} \qquad (i = 1, \ldots, n; s = 1, \ldots, k), \tag{2.36}$$

where x_{is} is the value of the explanatory variable for the ith individual in the sth set, α_s is a parameter for the sth set and β is the common regression coefficient on x. We consider in Section 2.7 the problem of examining the consistency of the data with such a model.

Under (2.36), the conditioning statistics are the separate sample numbers of successes $t_{0,1}, \ldots, t_{0,k}$ in the k sets, and the statistic associated with β is

$$T = \sum T_s, \tag{2.37}$$

where T_s is the test statistic associated with the sth set. Since the different sets are independent, the distribution of T, for any value of β, is the convolution of k distributions of the type investigated in Section 2.2.2. The adequacy of a normal approximation is usually improved by convolution.

If k is large it is sensible to make a graphical analysis of the separate T_s's. For instance, they can be separately standardized to have zero mean and unit variance and then plotted against relevant

See p. 34:
$T_s = \sum x_{is} Y_{is}$ is the total of a sample of size t_{0s} drawn from a finite population
$X_s = (x_{1s}, \ldots, x_{ns})$

properties of the groups. Also they can be ranked and plotted against expected normal order statistics.

As a special case we can deal with some problems of significance testing connected with regression on two variables, say x_1 and x_2. If for the ith individual the values of the explanatory variables are x_{1i} and x_{2i} a natural logistic model generalizing the simple regression model is

$$\lambda_i = \alpha + \beta_1 x_{1i} + \beta_2 x_{2i}. \qquad (2.38)$$

Unfortunately, if we are now interested in, say, β_1, it is not usually simple to find the appropriate conditional distribution. The following approach (Hitchcock, 1966) is often reasonable instead.

Group the x_2's into a fairly small number k of sets; of course the x_2's may already be so grouped. Then consider, instead of (2.38), the model in which for an individual in the sth set, i.e. having the sth value of x_2, with a value x_{1i} for x_1, the logistic transform is

$$\alpha_s + \beta x_{1i,s}. \qquad (2.39)$$

This is in one sense more general than (2.38), in that it does not postulate a linear regression on x_2; on the other hand, there may be some artificiality in grouping the values of x_2.

If the model (2.39) is a reasonable basis for the investigation of the partial regression on x_1, the analysis is based essentially on the discussion of Section 2.2.2. The same argument could be applied, for example, with three explanatory variables x_1, x_2 and x_3 provided that it is feasible to group the (x_2, x_3) combinations into a fairly small number of sets.

The procedure is illustrated by the following example.

Example 2.3 Analysis of ingot data (continued)
In Example 2.2 we analysed the dependence on x_1 ignoring x_2. A further analysis invoking a presumed linear logistic dependence on x_1 and x_2 is deferred to Section 2.6, but in the meantime we examine the dependence on x_1 exploiting the grouped form of x_2.

First consider possible regression on heating time, x_1. For each row of the table, i.e. for each fixed value of soaking time, x_2, we compute the test statistic for regression on x_1 and its null mean and variance. These are shown in Table 2.3. Thus in the first row the test statistic is

$$1 \times 27 + 3 \times 51 = 180,$$

$T_s = \Sum x_{i1} Y_{i1}$

i.e., the estimated regression slope $\hat{\beta}$.

and its null mean and variance are given by (2.30) with $t_0 = 4, n = 110$ and with m_1 and m_2 being the moments of the finite population of 110 values of x_1; in fact, $m_1 = 24.35$ and $m_2 = 145.07$. A similar calculation is made for the other rows and the combined test statistic is simply the sum of the separate test statistics, with null mean and variance given by the sum of the separate means and variances.

With more extensive data it would be useful to plot the standardized deviates in various ways to look for systematic features; thus these values might be plotted against the values of x_2 characterizing the rows and also the ranked deviates plotted against normal order statistics. In the present case, the general conclusion is clear. In four rows out of five, the test statistic exceeds its null expectation and the pooled test statistic gives highly significant evidence of an increase with x_1 in the probability of not being ready for rolling.

The corresponding analysis for regression on x_2 shows no evidence of departures from the null hypothesis.

Table 2.3 *Ingot data. Separate analyses for regression on* x_1 *and on* x_2
See p. 11
(a) Regression on x_1

Row	Test statistic	Null mean	Null st. dev.	Standardized deviate
1	180	97.4	23.8	3.48
2	108	74.7	16.5	2.02
3	28	36.4	11.7	−0.72
4	27	17.1	7.53	1.31
5	27	18.0	9.16	0.98
Pooled	370	243.7	33.4	3.79

(b) Regression on x_2

Col.	Test statistic	Null mean	Null st. dev.	Standardized deviate
1	—	—	—	—
2	4.4	4.32	1.27	0.06
3	14.6	13.32	2.41	0.53
4	3.0	3.92	1.24	−0.74
Pooled	22.0	21.56	2.99	−0.14

Soaking time x_2: 1.0, 1.7, 2.2, 2.8, 4.0

To summarize the data, the most informative quantities are probably the proportions not ready for rolling at each value of x_1, pooled over rows, i.e.

$$0/55 = 0, \quad 2/157 = 0.013, \quad 7/159 = 0.044, \quad 3/16 = 0.19.$$

These vary smoothly with x_1. A logistic regression equation could be fitted and this might be fruitful if, for example, several sets of data were involved and it was required to compare the dependences on x_1 in the different sets. With the present data, however, such fitting would normally add little information to that given directly by the above set of proportions.

As a less standard, but in some ways more elementary, example of regression analysis based on a logistic model we consider the testing of agreement between a series of binary observations and a set of numbers which purport to be the probabilities of success. Let $Y_1, \ldots, Y_n$ be independent binary random variables and let $p_1, \ldots, p_n$ be given constants, where $0 \leqslant p_i \leqslant 1$, the hypothesis to be examined being

$$\text{prob}(Y_i = 1) = p_i \quad (i = 1, \ldots, n).$$

Given a large amount of data, it would be possible to check adequacy by forming sub-groups with nearly constant p_i; the proportion of successes in each sub-group can then be compared with the relevant p_i. This method is not applicable with relatively small amounts of data.

One method of deriving a small sample test is to consider first a family of models derived by representing a translation on a logistic scale. Suppose in fact that the logistic transform for the ith trial is

$$\log\left(\frac{p_i}{1 - p_i}\right) + \alpha,$$

so that the corresponding probabilities of success and failure are

$$\frac{p_i e^\alpha}{p_i e^\alpha + 1 - p_i} \quad \text{and} \quad \frac{1 - p_i}{p_i e^\alpha + 1 - p_i}.$$

Thus the sufficient statistic is the total number of successes,

$$T_0 = \sum Y_i.$$

The probability-generating function of T_0 can be written down immediately. In particular, under the null hypothesis $\alpha = 0$,

$$E(T_0; \alpha = 0) = \sum p_i, \quad \text{var}(T_0; \alpha = 0) = \sum p_i(1 - p_i).$$

The distribution of T_0 can be approximated by a normal distribution with a continuity correction; in doubtful cases the skewness and kurtosis can be calculated or exact probabilities obtained.

The statistic T_0 examines only whether the p_i are systematically too high or too low; we also need to see whether they are too clustered or too dispersed. For this, suppose that the logistic transform for the ith trial is given by another single parameter model, namely

$$\beta \log\left(\frac{p_i}{1 - p_i}\right),$$

so that the probabilities of success and failure are respectively

$$\frac{p_i^\beta}{p_i^\beta + (1 - p_i)^\beta} \quad \text{and} \quad \frac{(1 - p_i)^\beta}{p_i^\beta + (1 - p_i)^\beta}.$$

The sufficient statistic for β can be obtained most conveniently by scoring for the ith observation

$$\begin{cases} \log(2p_i) & \text{if } Y_i = 1 \\ \log\{2(1 - p_i)\} & \text{if } Y_i = 0, \end{cases}$$

and by defining T_1 to be the total score. The factor 2 is included for convenience and to make events of probability $\frac{1}{2}$ score zero. The cumulants of T_1 can be written down and, when $\beta = 1$, the value for the null hypothesis is

$$E(T_1; \beta = 1) = n \log 2 + \sum p_i \log p_i + \sum (1 - p_i) \log(1 - p_i),$$
$$\text{var}(T_1; \beta = 1) = \sum p_i(1 - p_i)[\log\{p_i/(1 - p_i)\}]^2.$$

In practice we would examine both T_0 and T_1. Should we need to obtain a single test combining both statistics, the simplest procedure is to find the covariance matrix Ω of T_0 and T_1 from the above variances and the covariance of T_0 and T_1, namely

$$\sum p_i(1 - p_i) \log\{p_i/(1 - p_i)\}.$$

Then if $U = \{U_0, U_1\}$ is the row vector of differences from expectation, where $U_s = T_s - E(T_s; \alpha = 0, \beta = 1)$, then under the null hypothesis

$$U\Omega^{-1}U^{\mathrm{T}}$$

is distributed approximately as chi-squared with two degrees of freedom.

One possible application of the results of this section is in the testing of conformity between subjective probabilities, p_i, and realized events. Even here, it will usually be fruitful to enquire where the subjective probabilities come from, and on what evidence they are based, rather than merely to accept them as given. If there is a discrepancy between the subjective probabilities and the observations, it may be reasonable to assume that the 'objective' probability of an event whose subjective probability is p_i has a logistic transform given by a combination of the two models above, namely

$$\alpha + \beta \log\left(\frac{p_i}{1 - p_i}\right).$$

The parameters α and β can be estimated and the resulting equation used to convert the p_i's into estimated 'objective' probabilities.

2.3 2 × 2 Contingency table

2.3.1 Formulation

Example 1.1 illustrates the simplest possible type of comparison involving binary responses. There are just two groups of individuals, often corresponding to two treatments, labelled 0 and 1, and we aim to compare the probabilities of success in the two groups. Provided all individuals are independent and the probability of success is constant within each group, we are justified in condensing the data into a 2 × 2 table in the way summarized in Table 1.1. Only the numbers of trials and numbers of success in the two groups are relevant, n_s trials and R_s successes in group s ($s = 0, 1$).

We shall see later that the same configuration of data arises in two other ways, one in which the roles of rows and columns are interchanged (Section 4.3) and one in which rows and columns are treated symmetrically (Section 4.5), but for the time being we emphasize that rows and columns are treated quite differently as corresponding respectively to explanatory and response variables.

We use the general notation introduced earlier: ϕ_s is the probability of success in the sth group ($s = 0, 1$) and θ_i is the probability for the ith individual. Thus θ_i takes one of the two values ϕ_0, ϕ_1.

Our object is to compare ϕ_0 and ϕ_1. If the corresponding logistic transforms are λ_0 and λ_1, a linear logistic representation is

$$\lambda_0 = \alpha, \qquad \lambda_1 = \alpha + \Delta. \tag{2.40}$$

Here Δ represents the difference between groups on a logistic scale. Note that

$$\Delta = \lambda_1 - \lambda_0 = \log\left[\frac{\phi_1(1 - \phi_0)}{(1 - \phi_1)\phi_0}\right] = \log \psi, \tag{2.41}$$

say, where $\psi = e^\Delta$ is the ratio of the odds of success versus failure in the two groups.

Now (2.40) is a direct reparametrization of the problem from (ϕ_0, ϕ_1), taking values in the unit square, to (α, Δ), taking values in the whole plane. That is, no additional empirical assumptions about the data are involved beyond those already made in the representation with the pair of binomial parameters (ϕ_0, ϕ_1). It is convenient to say that the model (2.40) is saturated with parameters; that is, the number of independent parameters equals the number of independent binomial probabilities. The parametrization (2.40) is fruitful if Δ is in some sense a good way of characterizing the difference between the two groups.

Now there are many ways in which that difference might be measured. Thus if $h(\phi)$ is a monotonic increasing function of ϕ, we could take $h(\phi_1) - h(\phi_0)$, or some function of it, as such a measure. Possible choices of $h(\phi)$ are the following:

(a) $h(\phi) = \phi$, leading to $\phi_1 - \phi_0$;
(b) $h(\phi) = \log \phi$, leading to $\log(\phi_1/\phi_0)$ or to ϕ_1/ϕ_0;
(c) $h(\phi) = \log(1 - \phi)$, leading to $\log[(1 - \phi_1)/(1 - \phi_0)]$ or
     to $(1 - \phi_1)/(1 - \phi_0)$;
(d) $h(\phi) = \log[\phi/(1 - \phi)]$, leading to Δ of (2.41).

There are several criteria for choosing between alternative ways of assessing the difference between groups.

(α) Quite often it will be reasonable to require that if successes and failures are interchanged the measure of the difference between groups is either unaltered or at most changed in sign.

(β) It is desirable, wherever possible, to work with a measure that will remain stable over a range of conditions.

(γ) If a particular measure of difference has an especially clear-cut direct practical meaning, for example an explicit economic interpretation, we may decide to use it.

(δ) A measure of the difference for which the statistical theory of inference is simple is, other things being equal, a good thing.

Now consideration (α) excludes the simple ratio measures (b) and (c). Note, however, that when ϕ is very small, (d) is equivalent to (b), whereas when ϕ is very near 1, (d) is equivalent to (c).

Requirement (β) is particularly important. Ideally we would have the comparison of the two groups made under a range of circumstances, i.e. there would be a set of 2 × 2 tables in which the overall proportion of successes changed appreciably from set to set. It would then be an empirical matter to find that measure of difference, e.g. that function $h(\phi)$, if any, which is stable. By finding a scale on which the difference between groups is stable, there is economy in summarization and in addition the conclusions can be applied more confidently to a fresh set of conditions. In the absence of empirical data to establish the appropriate $h(\phi)$, it is worth noting that for the logistic model with a fixed Δ we can vary α in (2.40) arbitrarily and produce a family of situations in which the overall proportion of successes is arbitrary. On the other hand, in the corresponding representation directly in terms of probabilities, a given difference $\delta = \phi_1 - \phi_0$ is consistent with only a limited range of values for ϕ_1 and ϕ_0 individually. This makes it plausible, but not inevitable, that the parameter Δ in the logistic model (2.40) for the analysis of a single 2 × 2 table has a broader range of validity than the parameter $\delta = \phi_1 - \phi_0$.

Requirement (γ) may in some situations justify an analysis in terms of $\phi_1 - \phi_0$. If, say, the two groups represent alternative industrial processes and ϕ_1, ϕ_0 are the probabilities of defectives, then $m(\phi_1 - \phi_0)$ is approximately the difference between the numbers of defectives produced in a large number m of trials, and this may have an economic interpretation. Finally, while the requirement of simplicity of inference should not be of over-riding importance, we shall find that 'small-sample' theory of inference about the logistic parameter Δ is particularly simple.

If an analysis is done on a transformed scale it is important to present the conclusions in as meaningful a form as is feasible.

2.3.2 Logistic analysis

The comparison of the two probabilities on a logistic scale can be formulated in terms of the probabilities θ_i of success for individual i as

$$
\lambda = \begin{bmatrix} 1 \\ \vdots \\ 1 \\ 1 \\ \vdots \\ 1 \end{bmatrix} \alpha + \begin{bmatrix} 0 \\ \vdots \\ 0 \\ 1 \\ \vdots \\ 1 \end{bmatrix} \Delta, \tag{2.42}
$$

where the vectors of n elements are partitioned into sections of n_0 and n_1 components. The matrix x of the general model is here the $n \times 2$ matrix formed from the two columns in (2.42). In the general notation α corresponds to a nuisance parameter, and Δ to β, the parameter of interest.

Thus we have a very special case of the model for simple linear regression in which the vector of values of the single explanatory variable takes a particularly simple form and the discussion of Section 2.2 applies. Partly because of the intrinsic importance of the problem and more especially because the 2×2 table is the starting point for the discussion of a range of more complicated cases, we deal with this special case in some detail.

Thus the likelihood is

$$
\frac{e^{r_0 \alpha}}{(1 + e^\alpha)^{n_0}} \cdot \frac{e^{r_1(\alpha + \Delta)}}{(1 + e^{\alpha + \Delta})^{n_1}}. \tag{2.43}
$$

Note that the sufficient statistics are formed from the scalar product of the vector of binary observations with the vectors on the right-hand side of (2.42). Thus, in the general notation of Section 2.1,

$$
T_0 = R_. = \sum_{i=1}^{n_0 + n_1} Y_i = R_0 + R_1, \qquad T_1 = \sum_{i=n_0+1}^{n_0 + n_1} Y_i = R_1,
$$

where R_0 and R_1 are the numbers of successes in the two groups. It is convenient for the remainder of this section to use a notation tied to the special problem, rather than the general notation.

In accordance with the general discussion of Section 2.1, inference about Δ is based on the distribution of the random variable R_1 given that the observed value of $R_.$ is $r_.$. That is, we need the distribution of R_1, the number of successes in the group 1, given that the total number of successes in all is $r_.$.

To use the general formulae of Section 2.1.2, we first have to calculate the combinatorial coefficients in (2.5) and (2.6). These can be found from the generating function (2.6), which in this particular problem is

$$C(\zeta_0, \zeta_1) = (1 + \zeta_0)^{n_0}(1 + \zeta_0\zeta_1)^{n_1}. \tag{2.44}$$

Thus $c(r_., r_1)$, the coefficient of $\zeta_0^{r_.}\zeta_1^{r_1}$, is

$$c(r_., r_1) = \binom{n_0}{r_. - r_1}\binom{n_1}{r_1}, \tag{2.45}$$

and therefore from (2.9)

$$p_{R_1}(r_1; \Delta) = \frac{\binom{n_0}{r_. - r_1}\binom{n_1}{r_1} e^{\Delta r_1}}{\sum_u \binom{n_0}{r_. - u}\binom{n_1}{u} e^{\Delta u}}. \tag{2.46}$$

An important special case is obtained when $\Delta = 0$; then (2.46) becomes the hypergeometric distribution

$$p_{R_1}(r_1; 0) = \frac{\binom{n_0}{r_. - r_1}\binom{n_1}{r_1}}{\binom{n_0 + n_1}{r_.}}, \tag{2.47}$$

since the sum in the denominator of (2.46) can be given explicitly when $\Delta = 0$.

There is a direct argument for (2.47) which is in line with the more general context of Section 2.2. Given $r_.$, i.e. the total number of 1s, and that $\Delta = 0$, all distinct binary vectors with $r_.$ 1s and $n_0 + n_1 - r_.$ 0s have equal probability. Hence the number of 1s in the second sample of size n_1 can be regarded as the number of 1s in a sample of size n_1 drawn randomly without replacement from a finite population of $r_.$ 1s and $n_0 + n_1 - r_.$ 0s. It is well known that random sampling of this finite population without replacement leads to the hypergeometric distribution (2.47) (Feller, 1968, §II.6).

2.3.3 Significance test

While estimation is under most circumstances more important then testing significance, an important test of the null hypothesis $\Delta = 0$, Fisher's 'exact' test (Pearson and Hartley, 1966, Table 38), follows directly by computing tail areas from (2.47) via (2.16)–(2.19).

It follows from the known properties of the hypergeometric distribution, or from results on the sampling of general finite populations, that if $E(R_1;0)$ and var $(R_1;0)$ denote the mean and variance of (2.47), then

$$E(R_1;0) = n_1 r./(n_0 + n_1), \tag{2.48}$$

$$\text{var}(R_1;0) = \{n_0 n_1 r.(n_0 + n_1 - r.)\}/\{(n_0 + n_1)^2 (n_0 + n_1 - 1)\}.$$

An approximation to the tail areas required for the test can be obtained by using the standard normal integral with argument

$$\frac{|r_1 - E(R_1;0)| - \frac{1}{2}}{\sqrt{\text{var}(R_1;0)}}. \tag{2.49}$$

The statistic (2.49) differs very slightly from the square root of the usual chi-squared statistic, corrected for continuity (Pearson, 1947).

In fact $|r_1 - E(R_1;0)|$ is the absolute value of the difference between observed and fitted frequencies for any of the four cells. The fitted frequencies are

$$n_0 r./(n_0 + n_1), \qquad n_1 r./(n_0 + n_1),$$

$$n_0 (n_0 + n_1 - r.)/(n_0 + n_1), \qquad n_1 (n_0 + n_1 - r.)/(n_0 + n_1)$$

and the sum over the four cells of the reciprocals of the fitted frequencies is easily shown to be the reciprocal of var$(R_1;0)$ in (2.48) with $n_0 + n_1 - 1$ replaced by $n_0 + n_1$.

Example 2.4 Proportions of smokers in two groups of physicians (continued)
For numerical illustration we used the data of Table 1.2. The statistic (2.49) is

$$\frac{|3 - 8.32| - \frac{1}{2}}{\sqrt{2.957}} = 2.80,$$

corresponding to a one-sided significance level of 0.0026. The value obtained by summing hypergeometric probabilities to give the probability of a deviation as or more extreme than that observed is 0.0025.

A surprising variety of other procedures have been proposed for this problem. Although some, for example that based on the empirical logistic transform, can be useful as a basis for tackling more complicated problems, we can for the great majority of purposes see no point in using other than the simple procedure above.

2.3.4 Interval estimation

In principle confidence intervals are found from the non-central distribution (2.46), essentially by finding those values of Δ for which the observed value r_1 is just not significantly different from Δ at the level in question; see (2.18) and (2.19) for the general formulation.

This procedure is a little cumbersome for general use and impracticable for most more complicated problems of similar type. There are a variety of simpler approximations. Those of some interest, in particular as a basis for tackling more complicated problems, are in outline as follows:

1. via a normal approximation to the difference of two empirical logistic transforms;
2. almost equivalently, via maximum likelihood estimates derived from the unconditional likelihood function (2.43). Note that the maximum likelihood estimate of the probability of success in a group is the corresponding sample proportion and that therefore the unconditional maximum likelihood estimate of Δ is the difference of the two (unmodified) logistic transforms (2.21), i.e.

$$\hat{\Delta} = \log\left[\frac{R_1}{n_1 - R_1}\right] - \log\left[\frac{R_0}{n_0 - R_0}\right], \qquad (2.50)$$

with variance estimated, for example via the observed information matrix (A1.18), as equal to

$$\frac{n_0}{R_0(n_0 - R_0)} + \frac{n_1}{R_1(n_1 - R_1)}. \qquad (2.51)$$

The slightly modified forms based on (2.26) and (2.27) are likely to give a better result when some of the frequencies are small;

3. the unconditional likelihood functions (2.43) can be used to

derive a profile log likelihood for Δ following the prescription of (A1.52);

4. the likelihood-based arguments of (2) and (3), can be employed, using the conditional likelihood function (2.46) rather than the unconditional likelihood function (2.43). In particular, the conditional maximum likelihood estimate $\hat{\Delta}_c$ maximizes the conditional log likelihood

$$l_c(\Delta) = \Delta r_1 - \log\left\{ \sum_u \binom{n_0}{r_. - u}\binom{n_1}{u} e^{\Delta u} \right\} \qquad (2.52)$$

and, because of the exponential family form of (2.52), $\hat{\Delta}_c$ satisfies

$$r_1 = E(R_1 \mid R_. = r_.; \hat{\Delta}_c)$$
$$= \sum_u u \binom{n_0}{r_. - u}\binom{n_1}{u} e^{\hat{\Delta}_c u} \Bigg/ \sum_u \binom{n_0}{r_. - u}\binom{n}{u} e^{\hat{\Delta}_c u}. \qquad (2.53)$$

The conditional maximum likelihood estimate $\hat{\Delta}_c$ is not the same as the unconditional maximum likelihood estimate $\hat{\Delta}$; it is conjectured that $\hat{\Delta}_c$ is always nearer to zero than $\hat{\Delta}$; see Table 2.4 and Exercise 31.

An approximate $1 - \alpha$ confidence limit for Δ can be obtained from the conditional log likelihood function (2.52), via

$$\{\Delta;\ 2\{l_c(\hat{\Delta}_c) - l_c(\Delta)\} \leqslant \chi^2_{1,\alpha}\}, \qquad (2.54)$$

where $\chi^2_{1,\alpha}$ is the upper α point of the chi-squared distribution with one degree of freedom; note that $\chi^2_{1,\alpha} = k^{*2}_{1/2\alpha}$, where $k^*_{1/2\alpha}$ is the upper $\frac{1}{2}\alpha$ point of the standard normal distribution. Computation of $l_c(\Delta)$ is tedious unless one of n_0, n_1 is reasonably small. A reasonably simple and accurate approximation to the conditional log likelihood function based on a saddle-point expansion (Barndorff-Nielsen and Cox, 1979) is

$$l_c(\Delta) \simeq \tfrac{1}{2}\log\{n_0 \hat{\xi}_\Delta(1 + \hat{\xi}_\Delta)^{-2} + n_1 \hat{\xi}_\Delta e^\Delta (1 + e^\Delta \hat{\xi}_\Delta)^{-2}\}$$
$$\qquad - n_0 \log(1 + \hat{\xi}_\Delta) - n_1 \log(1 + e^\Delta \hat{\xi}_\Delta) + r_0 \log \hat{\xi} + r_1 \Delta, \qquad (2.55)$$

where $\hat{\xi}_\Delta$ is the positive root of the quadratic equation

$$\xi^2 e^\Delta(n_0 + n_1 - r_.) - \xi\{(n_0 - r_.) + (n_1 - r_.)e^\Delta\} - r_. = 0,$$

so that calculation of (2.55) essentially reduces to solving a quadratic equation.

Table 2.4 *Estimates of* Δ *for data of Table 1.2*

Method	95% confidence interval	Point estimate
Exact (conservative)	(0.47, 3.52)	—
Condtl log lik.	(0.67, 3.44)	1.91
Saddle point	(0.67, 3.44)	1.91
Emp. logit transform	(0.59, 3.03)	1.81
Uncontl log lik.	(0.58, 3.27)	1.93

Example 2.5 Proportion of smokers in two groups of physicians (continued)
The close agreement of the various methods of calculating confidence limits and indeed point estimates for Δ is illustrated by Table 2.4.

Of course when confidence intervals are so wide, precision in their calculation is unnecessary, but the arguments involved are important as a basis for handling more complicated problems.

2.3.5 Difference of probabilities

While the main emphasis of this chapter is on interpretations based on the logistic model, it is important to discuss other possible formulations. As noted in Section 2.2.1 we may compare ϕ_1 and ϕ_0 on other scales than the logistic, the simplest and most important being via the simple difference $\delta = \phi_1 - \phi_0$. Of course testing $\delta = 0$ is entirely equivalent to testing $\Delta = 0$, but as soon as we come to estimation a different approach is needed. There is no 'exact' conditional treatment; the simplest procedure is to assign a large-sample standard error to the maximum likelihood estimate $\hat{\delta} = R_1/n_1 - R_0/n_0$, but for the reasons sketched in Section A1.5 the more satisfactory approach is via the profile log likelihood for δ.

To calculate this, write $\phi_0 = \phi - \frac{1}{2}\delta$, $\phi_1 = \phi + \frac{1}{2}\delta$ so that the unconditional log likelihood is

$$l(\delta, \phi) = (n_0 - r_0)\log(1 - \phi + \tfrac{1}{2}\delta) + r_0 \log(\phi - \tfrac{1}{2}\delta)$$
$$+ (n_1 - r_1)\log(1 - \phi - \tfrac{1}{2}\delta) + r_1 \log(\phi + \tfrac{1}{2}\delta).$$

The overall maximum is at

$$\hat{\phi} = \tfrac{1}{2}(R_0/n_0 + R_1/n_1), \quad \hat{\delta} = (R_1/n_1 - R_0/n_0),$$

Cox, D. R and E. Snell, Analysis of Binary Data,
2nd ed., 1989

whereas for fixed δ, the maximum likelihood estimate of ϕ, $\hat{\phi}_\delta$, satisfies

$$-\frac{(n_0 - r_0)}{(1 - \hat{\phi}_\delta + \frac{1}{2}\delta)} + \frac{r_0}{(\hat{\phi}_\delta - \frac{1}{2}\delta)} - \frac{(n_1 - r_1)}{(1 - \hat{\phi}_\delta - \frac{1}{2}\delta)} + \frac{r_1}{(\hat{\phi}_\delta + \frac{1}{2}\delta)} = 0,$$

$$(2.56)$$

a quadratic equation in $\hat{\phi}_\delta$. The profile log likelihood for δ is $l(\delta, \hat{\phi}_\delta)$ with its maximum at $\delta = \hat{\delta}$. Approximate confidence intervals for δ are obtained in the standard way from

$$\{\delta; 2[l(\hat{\delta}, \hat{\phi}) - l(\delta, \hat{\phi}_\delta)] \leqslant \chi^2_{1,\alpha}\}.$$

2.4 Matched pairs

Suppose that individuals are paired and that in each pair one individual is assigned at random to treatment 0, the other to treatment 1. On each individual a binary response is observed. The possible responses on a pair of individuals are thus 00, 10, 01 and 11, the response to treatment 0 being written first. Let $Y_{0,s}$ and $Y_{1,s}$ represent the responses on the two individuals forming the sth pair.

Let the numbers of pairs with the four types of response be R^{00}, R^{10}, R^{01} and R^{11}. Then $\sum R^{lm}$ is equal to the number of pairs, say k, the total number of binary responses being $n = 2k$. Paired data can, of course, arise also from observational rather than from experimental studies.

The analysis depends appreciably on the model thought to be appropriate. Thus if it were assumed provisionally that all individuals respond independently with probabilities of success θ_0 and θ_1, for treatments 0 and 1, then by arguments of sufficiency only the total numbers of successes $R^{10} + R^{11}$ and $R^{01} + R^{11}$ in the two treatment groups need be considered. They can be compared by the procedure of Section 2.3. Such an analysis would, however, ignore the correlation between the two individuals in a pair.

Suppose, then, to represent the pairing we adopt a linear logistic model that is analogous to the normal-theory linear model commonly used for paired comparison quantitative data. Then, for the sth pair, the logistic transforms for treatments 0 and 1 are respectively

$$\alpha_s \quad \text{and} \quad \alpha_s + \Delta, \tag{2.57}$$

where α_s is a nuisance parameter characteristic of the sth pair, and Δ is a treatment effect assumed constant on the logistic scale.

constant treatment effect assumed on the logistic scale

In the general analysis of Section 2.1.3, the statistics associated with the nuisance parameters, and hence used for conditioning, are the pair totals $Y_{0,s} + Y_{1,s}$ $(s = 1, \ldots, k)$. The statistic associated with the parameter Δ is the total number of successes for treatment 1, i.e. $T = R^{01} + R^{11}$. To examine the conditional distribution of T given $Y_{0,s} + Y_{1,s}$ $(s = 1, \ldots, k)$, we argue as follows. Any pair for which $Y_{0,s} + Y_{1,s} = 0$ must contribute zero to T. Any pair for which $Y_{0,s} + Y_{1,s} = 2$ must contribute one to T. Hence only pairs for which $Y_{0,s} + Y_{1,s} = 1$, i.e. only the pairs with 'mixed' responses 01 and 10, contribute to T an amount to be regarded as random. Thus the conditional distribution required is in effect that of the number R^{01} of pairs 01, given that $R^{01} + R^{10} = m$, the total number of 'mixed' pairs.

Under (2.57) the conditional probability that the sth pair contributes one to R^{01}, given that it is 'mixed', is

$$\text{prob } (Y_{0,s} = 0, Y_{1,s} = 1 \,|\, Y_{0,s} + Y_{1,s} = 1)$$

$$= \frac{\left(\dfrac{1}{1 + e^{\alpha_s}}\right)\left(\dfrac{e^{\alpha_s + \Delta}}{1 + e^{\alpha_s + \Delta}}\right)}{\left(\dfrac{1}{1 + e^{\alpha_s}}\right)\left(\dfrac{e^{\alpha_s + \Delta}}{1 + e^{\alpha_s + \Delta}}\right) + \left(\dfrac{e^{\alpha_s}}{1 + e^{\alpha_s}}\right)\left(\dfrac{1}{1 + e^{\alpha_s + \Delta}}\right)}$$

$$= \frac{e^{\Delta}}{e^{\Delta} + 1}. \tag{2.58}$$

Because this is the same for all pairs, it follows that the conditional distribution of R^{01} is binomial with index m and parameter (2.58). In particular, under a null hypothesis $\Delta = 0$, the binomial parameter equals $\frac{1}{2}$.

binomial test

It is thus possible to test hypotheses about, or to obtain confidence limits for, Δ, using methods for a single binomial sample. Further, given several independent sets of data of the above form, each with its appropriate Δ, we may set up a linear logistic model for the Δ's and then use some of the other special techniques given in this book.

NB

Example 2.6 Some psychiatric paired comparison data
Maxwell (1961, p. 28) has given some data on twenty-three matched pairs of depressed patients, one member of each pair being classed as 'depersonalized', 0, the other as 'not depersonalized', 1. After treatment each patient is classified as 'recovered', coded as $Y = 1$, or 'not recovered', coded as $Y = 0$. Table 2.5 summarizes the results. These data are convenient for simple illustrative purposes.

Table 2.5 *Recovery of twenty-three pairs of patients*

	0 Depersonalized	1 Not depersonalized	No. of pairs
		Response	
	0	0	2
	1	0	2
	0	1	5
	1	1	14

To test the difference between groups, taking as the null hypothesis $\Delta = 0$, we consider the seven 'mixed' pairs and ask whether the split into two and five is consistent with binomial sampling with seven trials and parameter $\frac{1}{2}$. Note that, in particular, the fourteen pairs in which both patients recover are disregarded. The effective numbers of observations are very small and in practice a formal test would not be necessary. The 'exact' two-sided level of significance is, however,

$$2\left[\binom{7}{0}+\binom{7}{1}+\binom{7}{2}\right]\frac{1}{2^7}=\frac{29}{64}.$$

[margin: limitation of stratified logistic model (conditional) ML estimation]

The pairs giving responses 00 or 11 are ignored because in (2.57) each pair has an arbitrary associated parameter α_s and there is the very strong requirement that the probability properties of the method of analysis must be the same whatever the values of $\alpha_1, \ldots, \alpha_k$. Thus a pair giving a response 00 might have had a very large negative α_s disguising the presence of a treatment effect Δ. If, however, some restriction is placed on the variation of the α_s's, some relevant information may be contained in the numbers of 00 and 11 pairs. In fact the occurrence of a large number of such pairs is often evidence that Δ is small.

[margin: adumbrates hierarchical models]

If, for example, each pair is characterized by one or more explanatory variables, a possible approach is to replace α_s by a function, probably linear, of those explanatory variables and to fit the resulting model, which would now contain a much smaller number of unknown parameters. An alternative approach is to group the pairs into sets with the same or similar values of the explanatory variables and to use the analysis for the combination of several 2×2 tables. Both these analyses assume that the explanatory variables have 'accounted for' all or most of the correlation present.

A different approach paralleling the model (2.57), i.e. not requiring the availability of explanatory variables, is to regard each ~~individual~~ *pair* as yielding one of the four possible responses $(0,0), (0,1), (1,0), (1,1)$, the corresponding probabilities being

$$\theta_{00}, \theta_{01}, \theta_{10}, \theta_{11}, \quad \sum \theta_{ij} = 1. \tag{2.59}$$

multinomial model

Now the probabilities of success under the two treatments are respectively $\theta_{01} + \theta_{11}$ and $\theta_{10} + \theta_{11}$. The equality of these, allowing for arbitrary dependency, is equivalent to testing $\theta_{01} = \theta_{10}$ and, so long as θ_{11} (and θ_{00}) are arbitrary we are led to the test for $\Delta = 0$ in (2.58).

The difference between the models emerges when we consider estimation. The model (2.57) leads to the estimation of Δ via (2.58), which can be interpreted as the logit difference for an arbitrary individual. Under the multinomial model (2.59), however, it would seem more natural to estimate some contrast of the probabilities of success in the two treatment groups, and on a direct or logistic scale this would be respectively

direct scale   *logistic scale*

$$\theta_{10} - \theta_{01}, \quad \log\left(\frac{\theta_{10} + \theta_{11}}{\theta_{01} + \theta_{00}}\right) - \log\left(\frac{\theta_{01} + \theta_{11}}{\theta_{01} + \theta_{00}}\right). \tag{2.60}$$

$$= \log\left(\frac{\theta_{10} + \theta_{11}}{\theta_{01} + \theta_{11}}\right)$$

Confidence intervals for the parameters (2.60) are preferably calculated via a profile log likelihood for the parameter in question; this involves all the data and not solely the 'mixed' pairs. A simpler approach is via the maximum likelihood estimates and their large sample variances which are respectively

$$(R^{10} - R^{01})/k, \quad (R^{10} + R^{01})/k^2$$

for the difference and for the logistic difference

$$\log(R^{10} + R^{11}) - \log(R^{01} + R^{00}) - \log(R^{01} + R^{11}) + \log(R^{10} + R^{00}),$$
$$(R^{01} - R^{10})^2 \{R^{00}(R^{00} + R^{01})^{-2}(R^{00} + R^{10})^{-2}$$
$$+ R^{11}(R^{10} + R^{11})^{-2}(R^{01} + R^{11})^{-2}\}$$
$$+ R^{01}(k + R^{01} - R^{10})^2(R^{00} + R^{01})^{-2}(R^{01} + R^{11})^{-2}$$
$$+ R^{10}(k + R^{10} - R^{01})^2(R^{10} + R^{11})^{-2}(R^{00} + R^{10}). \tag{2.61}$$

Again all observations are used.

For most practical purposes the simple analysis based on the rather general model (2.57) is likely to be preferred.

A serious criticism of the model is, however, that there is no check from the data on its adequacy. Note, though, that as a test of significance of $\Delta = 0$, the binomial test has the correct probability

$$\text{Var}\left(\frac{R^{10} - R^{01}}{k}\right) = \frac{1}{k^2}\left\{\text{Var}(R^{10}) + \text{Var}(R^{01}) - 2\,\text{Cov}(R^{10}, R^{01})\right\}$$

properties whenever there is no treatment effect; the model (2.57) serves to indicate one special set of circumstances under which rejecting the pairs giving responses 00 or 11 is the efficient thing to do. An indirect justification of the model would be obtained if we had several sets of data comparing the two treatments under different conditions and with different overall success rates, the data being consistent with a single constant Δ.

2.5 Several 2×2 contingency tables

2.5.1 Formulation

In the problem treated in Section 2.4, the observations on each pair of individuals constitute a rather degenerate form of 2×2 contingency table, in which in the notation of Table 1.1, $n_0 = n_1 = 1$, i.e. there is one observation in each group or on each treatment. A natural and important generalization of the problem of Section 2.4 is thus to consider a set of k general 2×2 contingency tables, all involving a comparison of the same two treatments; see Example 1.2.

Quite often the k tables will have been obtained by subdividing data initially in the form of a single 2×2 table; the dangers of not partitioning the data in this way are illustrated in extreme form in Section 2.8.

Suppose that in the sth table $R_{0,s}$ and $R_{1,s}$ are the numbers of successes in the two groups, the corresponding sample sizes being $n_{0,s}$ and $n_{1,s}$.

Essentially two types of problem arise. The first is concerned with the possible homogeneity of the group (or treatment) differences in the separate tables. For example, are the different tables reasonably consistent with a treatment effect constant on the logistic scale? Secondly, we may tentatively assume this constant effect and wish to obtain confidence limits for it and, in particular, to test the significance of the effect, combining the evidence from the separate tables.

2.5.2 Significance test

The first type of problem will be deferred to Section 2.5.5, where we deal with methods in which several parameters are of simultaneous interest. Suppose, then, that we tentatively take a model with a constant effect on the logistic scale; we allow arbitrary differences between tables in, say, the probabilities in group 1. For the sth table we take the logistic transforms of the probabilities of success in groups 0 and 1 to be respectively

$$\alpha_s \quad \text{and} \quad \alpha_s + \Delta.$$

We can now either appeal to the general results of Section 2.1 or proceed again from first principles to write down the combined likelihood of all observations. It is found that inference about Δ is based on the conditional distribution of

$$T = R_{1,\cdot} = \sum R_{1,s}, \tag{2.62}$$

given the marginal totals of all tables.

Now the distribution of $R_{1,s}$ for a particular s is given by the generalized hypergeometric distribution (2.46) and the required distribution of T is the convolution of k of these distributions. It is clear that, except in some simple cases, notably that of Section 2.4, this is impracticable for exact calculation of confidence limits. However, we can test the null hypothesis that $\Delta = 0$ by noting that from (2.48) the mean and variance of T are

$$E(T; 0) = \sum \frac{n_{1,s} r_{\cdot,s}}{n_{0,s} + n_{1,s}}, \tag{2.63}$$

$$\text{var}(T; 0) = \sum \frac{n_{0,s} n_{1,s}(n_{0,s} + n_{1,s} - r_{\cdot,s}) r_{\cdot,s}}{(n_{0,s} + n_{1,s})^2 (n_{0,s} + n_{1,s} - 1)}, \tag{2.64}$$

where $r_{\cdot,s}$ is the observed total number of successes in the sth table. A normal approximation, with continuity correction, for the distribution of T will nearly always be adequate; the approximation is good even for a single table and will be improved by convolution. In exceptional cases a correction based on the third and fourth cumulants can be introduced.

This procedure leads to a combined test of significance from several independent 2×2 contingency tables. It is preferable to tests sometimes proposed based on the addition of chi or chi-squared values, because allowance is made for the differing amounts of information in the separate tables. However, if the tables show effects in opposite directions or if, say, only one of the tables has an effect, the use of T will not be effective. In practice these possibilities can never be excluded and it will be a wise precaution to supplement (2.62), for example by graphical analyses of the separate estimates of Δ based on the empirical logistic transform; see Section 2.1.6.

2.5.3 A simple example

Example 1.2 is special in that $n_{1,s} = 1$ for all s; it is intermediate between the matched pair problem and the general case in which $n_{0,s}$ and $n_{1,s}$ are not one. In Example 1.2 the contribution of the sth set to the combined statistic T is either 0 or 1 and under the null hypothesis the probability generating function of the contribution is

$$\frac{n_{0,s} + 1 - r_{\cdot s}}{n_{0,s} + 1} + \frac{r_{\cdot s}}{n_{0,s} + 1}\, \zeta, \qquad (2.65)$$

where $r_{\cdot s}$ is the observed value of the conditioning statistic $R_{0,s} + R_{1,s}$ and ζ is the argument of the probability generating function. Thus the probability generating function of T under the null hypothesis is

$$\prod \left(\frac{n_{0,s} + 1 - r_{\cdot,s}}{n_{0,s} + 1} + \frac{r_{\cdot,s}}{n_{0,s} + 1}\, \zeta \right). \qquad (2.66)$$

For general Δ, (2.66) is replaced by

$$\prod \{ 1 - \theta_s(\Delta) + \theta_s(\Delta)\zeta \}, \qquad (2.67)$$

where

$$\theta_s(\Delta) = \frac{r_{\cdot,s}\, e^{\Delta}}{r_{\cdot,s}\, e^{\Delta} + n_{0,s} - r_{\cdot,s} + 1}$$

is the conditional probability that $R_{1,s} = 1$, given that $R_{0,s} + R_{1,s} = r_{\cdot,s}$.

Example 2.7 The crying of babies

We can apply these results to the data of Table 1.3, where the treatments 0 and 1 refer to control and experimental babies and the special condition $n_{1,s} = 1$ holds, there being only one experimental baby on each day. The test statistic T, the total number of successes in group 1, has the observed value 15 and, from (2.63) and (2.64)

$$E(T; 0) = 11.47, \qquad \text{var}\,(T; 0) = 3.420.$$

Thus the standardized deviate, corrected for continuity, is $3.03/\sqrt{3.420} = 1.64$, corresponding to significance at the 5% level in a one-sided normal test. It is fairly clear that in this example the null distribution of T will be negatively skew and hence that the normal approximation will give too large a value. As a check on this the coefficients of ζ^{18}, ζ^{17}, ζ^{16} and ζ^{15} in the generating function (2.66) were evaluated recursively by computer. The 'exact' significance level, the sum of these coefficients, is 0.045.

Cox (1966a) has considered this and other ways in which these data might be analysed.

2.5.4 Interval estimation

We have seen that there are a number of methods of interval estimation for a single 2×2 table. It is no surprise, therefore, that there is a profusion of methods for several 2×2 tables. We shall not attempt a complete review. An important general point is that if k, the number of tables, is relatively large and the individual tables relatively small, straightforward use of unconditional maximum likelihood can be quite misleading. In particular if $n_{0,s} = n_{1,s} = 1$ $(s = 1, \ldots, k)$, then for large k, the unconditional maximum likelihood estimate $\hat{\Delta}$ is close to 2Δ, not to Δ; see Exercise 32.

On general grounds, under the model in which the logistic transforms in the sth table are α_s and $\alpha_s + \Delta$, where $\alpha_1, \ldots, \alpha_k$ are totally arbitrary, the basis for inference about Δ is the conditional log likelihood derived via (2.52) as

$$l_c(\Delta) = \Delta \sum_s r_{1,s} - \sum_s \log \left\{ \sum_u \binom{n_{0,s}}{r_{.,s} - u} \binom{n_{1,s}}{u} e^{\Delta u} \right\}. \qquad (2.68)$$

From this the conditional maximum likelihood estimate can be determined. It satisfies

$$\sum_s r_{1,s} = E\left(\sum_s R_{1,s} \mid R_{.,s} = r_{.,s}(s = 1, \ldots, k); \hat{\Delta}_c \right). \qquad (2.69)$$

Confidence limits can be obtained approximately in the usual way by substituting (2.68) into (2.54). The calculations are much simplified by use of the saddle-point approximation (2.55) applied separately to each 2×2 table.

2.5.5 Some simpler procedures

There are broadly three different routes to confidence limits for Δ that are useful if the computations involved in (2.68) are judged too arduous. These are

(a)   to obtain an estimate of Δ from the sth table as

$$\tilde{\Delta}_s^{(w)} = \log\left(\frac{R_{1,s} - \frac{1}{2}}{n_{1,s} - R_{1,s} - \frac{1}{2}}\right) - \log\left(\frac{R_{0,s} - \frac{1}{2}}{n_{0,s} - R_{0,s} - \frac{1}{2}}\right) \quad (2.70)$$

with estimated variance

$$\tilde{v}_s^{(w)} = \frac{(n_{1,s} - 1)}{R_{1,s}(n_{1,s} - R_{1,s})} + \frac{(n_{0,s} - 1)}{R_{0,s}(n_{0,s} - R_{0,s+1})}, \quad (2.71)$$

obtained via (2.28), this form of empirical transform being chosen because of the use as an overall estimate of Δ of

$$\tilde{\Delta} = \left(\sum \tilde{\Delta}_s^{(w)}/\tilde{v}_s^{(w)}\right)\left(\sum 1/\tilde{v}_s^{(w)}\right)^{-1} \quad (2.72)$$

with variance $\left(\sum 1/\tilde{v}_s^{(w)}\right)^{-1}$.

(b)   to use unconditional maximum likelihood, that is to work directly from

$$\begin{aligned}
l(\alpha_1, \ldots, \alpha_k, \Delta) = &\sum \alpha_s r_{\cdot,s} + \Delta \sum r_{1,s} \\
&- \sum n_{0,s} \log(1 + e^{\alpha_s}) \\
&- \sum n_{1,s} \log(1 + e^{\alpha_s + \Delta}).
\end{aligned} \quad (2.73)$$

This can be maximized numerically, for example by any of several library packages, and confidence limits for Δ obtained either via a standard error computed by inverting the observed information matrix or, preferably, via a profile log likelihood function for Δ.

(c)   to use a simple estimate of Mantel and Haenszel (1959), namely

$$\begin{aligned}
\tilde{\Delta}_{\text{MH}} = &\log\left[\sum R_{1,s}(n_{0,s} - R_{0,s})/(n_{0,s} + n_{1,s})\right] \\
&- \log\left[\sum (n_{1,s} - R_{1,s})R_{0,s}/(n_{0,s} + n_{1,s})\right]. \quad (2.74)
\end{aligned}$$

Note that a single term of (2.74) gives essentially $\tilde{\Delta}$, so that $\tilde{\Delta}_{\text{MH}}$ can be regarded as a special form of weighted means, based on a linearizing approximation near $\Delta = 0$.

This equivalence can be used to motivate various estimates of the variance of $\tilde{\Delta}_{\text{MH}}$. These have been compared by simulation by Robins, Breslow and Greenland (1986); see also Donner and Hauck (1988). Robins et al. find that an estimated variance performing well in their simulations is

$$\tfrac{1}{2}\{\sum A_s C_s/C^2 + \sum(A_s D_s + B_s C_s)/(C.D.) + \sum B_s D_s/D^2\},$$

where

$$A_s = (R_{0,s} + n_{1,s} - R_{1,s})/(n_{0,s} + n_{1,s}),$$
$$B_s = (R_{1,s} + n_{0,s} - R_{0,s})/(n_{0,s} + n_{1,s}),$$
$$C_s = R_{0,s}(n_{1,s} - R_{1,s})/(n_{0,s} + n_{1,s}),$$
$$D_s = R_{1,s}(n_{0,s} - R_{0,s})/(n_{0,s} + n_{1,s})$$

and

$$C. = \sum C_s, \quad D. = \sum D_s.$$

Example 2.8 The crying of babies (continued)
This example provides a convenient if rather extreme base for comparing the above methods. The estimate $\hat{\Delta}_c$, its saddle point approximation, the unconditional maximum likelihood estimate $\hat{\Delta}$ and the Mantel–Haenszel estimate are 1.26, 1.27, 1.43 and 1.20. The corresponding 95% confidence intervals via the first two methods are respectively (0.04, 2.80) and (0.04, 2.82).

2.5.6 Test of homogeneity

The above discussion is on the basis of an assumption that the logistic difference is constant across tables. If there are no or few tables with very small numbers of 0s and 1s, method (a) of Section 2.5.5, using the empirical logistic transform, is very convenient for assessing homogeneity, i.e. constancy of Δ. The statistic

$$\sum(\tilde{\Delta}_s^{(w)} - \tilde{\Delta})^2/\tilde{v}_s^{(w)}$$

has approximately a chi-squared distribution with $k - 1$ degrees of freedom and the individual residuals $(\tilde{\Delta}_s^{(w)} - \tilde{\Delta})/\sqrt{\tilde{v}_s^{(w)}}$ can be analysed approximately as normal-theory residuals of zero mean and unit variance.

An alternative procedure, reasonable when the individual tables are not small, is to fit by unconditional maximum likelihood

1. the above model with constant Δ;
2. the saturated model in which each table has its own logistic difference Δ_s, so that each table has two separate and unrelated probabilities of success, $2k$ parameters in all.

Comparison of the maximized log likelihood under (1) and (2) yields a test statistic with asymptotically a chi-squared distribution with $2k - (k + 1) = k - 1$ degrees of freedom.

Neither of these methods is satisfactory for large numbers of small tables, which in principle requires the use of conditional arguments. For this a simple reasonably practicable procedure is to construct a likelihood ratio test from the conditional log likelihood computed via the saddle-point approximation.

Exemplification of these points is postponed until after the discussion of residuals in Section 2.7.

2.5.7 Adjustment for a concomitant variable in a 2 × 2 table

Suppose that it is required to compare the probabilities of success for two groups of individuals and that for each individual a concomitant variable x is available, suspected of affecting the probability of success. We may require to compare the groups, adjusting for any differences associated with the concomitant variable. The situation is analogous to that treated by analysis of covariance in normal theory and is subject to the same broad conditions for its validity.

Sometimes, for instance in certain medical applications, an estimate of the average difference in probabilities of success may be required, the average being taken over a pre-assigned distribution of the concomitant variable, x. Thus x may be age, and probabilities of death may be standardized with respect to a 'standard' age distribution. Then the simplest procedure, at least with extensive data, is to divide the data into sets on the basis of a grouping of x. An estimated difference between groups can then be found for each set, and the weighted mean difference and its standard error found.

If an analysis in terms of a constant difference on a logistic scale is attempted, two approaches are possible, corresponding in fact to the two approaches to the regression problem contrasted in (2.38) and (2.39). If the values of x are grouped, we have the problem of Sections 2.5.1 and 2.5.2. This corresponds also to the analysis of (2.39). An alternative is also to assume logistic regression on x, the simplest assumption being to have the same slope in the two groups. This leads to the model

$$\lambda_i = \begin{cases} \alpha + \beta x_i & \text{in group 0,} \\ \alpha + \beta x_i + \Delta & \text{in group 1,} \end{cases}$$

exactly corresponding to the standard assumptions of linearity and parallelism involved in the simplest form of analysis of covariance.

The general procedures of this chapter now lead to an analysis for β, in particular to a simple test of the null hypothesis $\beta = 0$, and to a not so simple analysis for Δ, the parameter of main interest. For more complicated analyses, unconditional maximum likelihood can be used.

Example 2.9 Trend in a 2 × 2 contingency table
Consider an experiment to compare two treatments 0 and 1 which are tested in serial order in time. Table 2.6 gives some artificial data from such an experiment.

First, a test of a trend with serial order is a direct application of the results of Section 2.5. We calculate separately for treatments 0 and 1 the test statistics for trend, together with their null means and variances. The test statistics are

for 0: 195, with null expectation 173.0 and variance 192.1,

for 1: 154, with null expectation 122.2 and variance 354.4.

For example, the statistic for treatment 0 is the sum of the serial numbers of the successes in that group, $4 + 8 + \cdots + 28$. This is regarded as a random sample of size 11 from the finite population $\{2, 4, 7, \ldots, 28\}$.

The pooled test statistic is thus 349 leading to a standardized normal deviate, with continuity correction, of

$$\frac{|349 - 295.2| - \frac{1}{2}}{\sqrt{546.5}} = 2.29.$$

For both treatments the test statistic exceeds its null expectation and the pooled statistic, significant at about the 2% level, indicates quite strong evidence that the trend apparent from the inspection of the data is unlikely to be spurious.

The formulation of a corresponding procedure for the estimation of the difference between treatments, allowing for regression on serial number, is more complicated. The parameter of interest is now Δ and the conditioning statistics are the total number t_0 of successes, where in fact $t_0 = 19$; and the sum t_1 of the serial numbers over the successes, where in fact $t_1 = 349$. The test statistic is the number t_2 of successes under treatment 1. In principle we thus require to find $c(t_0, t_1, t_2)$, the number of distinct binary sequences with the relevant values of t_0, t_1

Table 2.6 *Comparison of two treatments tested in serial order*

Serial no., i	Treatment	Response	$\hat{\theta}_i$	d_i
1	1	0	0.177	−0.46
2	0	0	0.387	−0.80
3	1	1	0.215	1.91
4	0	1	0.446	1.11
5	1	0	0.258	−0.59
6	1	0	0.282	−0.63
7	0	0	0.537	−1.08
8	0	1	0.567	0.87
9	1	0	0.361	−0.75
10	1	1	0.390	1.25
11	0	1	0.653	0.73
12	1	0	0.449	−0.90
13	1	1	0.479	1.04
14	0	1	0.730	0.61
15	0	0	0.753	−1.75
16	0	1	0.775	0.54
17	0	0	0.795	−1.97
18	1	0	0.627	−1.30
19	1	1	0.655	0.73
20	0	1	0.848	0.42
21	0	1	0.863	0.40
22	0	1	0.877	0.37
23	1	1	0.755	0.57
24	1	0	0.777	−1.87
25	0	1	0.911	0.31
26	0	1	0.920	0.29
27	1	1	0.833	0.45
28	0	1	0.936	0.26
29	1	1	0.864	0.40
30	1	1	0.878	0.37

The fitted probabilities, $\hat{\theta}_i$, and the residuals, d_i, are discussed in Section 2.7.

and t_2; in fact $t_0 = 19$, $t_1 = 349$ and also t_2 varies over its full range of possible values, namely $4, \ldots, 15$.

Direct enumeration of these coefficients would be a formidable task and while it is possible to obtain an approximation much the simplest approach is to apply maximum likelihood ideas.

A rather systematic approach entirely by maximum likelihood methods is to fit five linear logistic models in which for the ith observation

$$\text{Model I:} \quad \lambda_i = \mu; \tag{2.75}$$

$$\text{Model II:} \quad \lambda_i = \begin{cases} \mu - \delta \text{ for treatment 0,} \\ \mu + \delta \text{ for treatment 1;} \end{cases} \tag{2.76}$$

$$\text{Model III:} \quad \lambda_i = \mu + \beta(i - 15); \tag{2.77}$$

$$\text{Model IV:} \quad \lambda_i = \begin{cases} \mu - \delta + \beta(i - 15) \text{ for treatment 0,} \\ \mu + \delta + \beta(i - 15) \text{ for treatment 1;} \end{cases} \tag{2.78}$$

$$\text{Model V:} \quad \lambda_i = \begin{cases} \mu - \delta + \beta(i - 15) \text{ for treatment 0,} \\ \mu + \delta + \gamma(i - 15) \text{ for treatment 1.} \end{cases} \tag{2.79}$$

Note that for computational reasons trends with serial number are expressed in terms of $(i - 15)$ rather than in terms of i, and the logistic difference $\Delta = 2\delta$ is expressed symmetrically in terms of δ. Verbal description of models I–V is unnecessary; the last model is included to cover the possibility of interaction, on a logistic scale, between treatments and serial order, although the possibility of detecting this with the present data is slight.

Table 2.7 summarizes the results of fitting the above models. The adequacy of the different models could be assessed from the significance of additional parameters, but is most neatly seen from the maximized log likelihoods, recalling that when one extra parameter is fitted the significance of the increase in maximized log likelihood is

Table 2.7 *Fitting of five models by maximum likelihood to the data of Table 2.6*

Model	No. of parameters	Maximized log likelihood	Selected estimates and asymptotic standard errors
I	1	−19.71	
II	2	−19.06	$\hat{\delta} = -0.440 \pm 0.39$
III	2	−16.66	$\hat{\beta} = 0.121 \pm 0.053$
IV	3	−16.04	$\hat{\delta} = -0.480 \pm 0.43;$ $\hat{\beta} = 0.121 \pm 0.055$
V	4	−16.00	$\hat{\delta} = -0.505 \pm 0.45;$ $\hat{\beta} = 0.139 \pm 0.092;$ $\hat{\gamma} = 0.111 \pm 0.068$

tested as one-half chi-squared with one degree of freedom. For example, in passing from model III to model IV one extra parameter, δ, is introduced and the increase in maximized log likelihood, 0.62, corresponds to a value of chi-squared of 1.24. Alternatively, the square of the ratio of $\hat{\delta}$ to its standard error is 1.25, in close agreement. Other examples of similar close equivalences can be found in the table.

There is no evidence of non-parallelism of the logistic regression lines for 0 and 1, and some evidence that there is trend with serial order. The estimated logistic difference between 1 and 0, $\hat{\Delta} = 2\hat{\delta}$, is slightly greater after adjustment for serial order than before. While well short of statistical significance at an interesting level, the estimated difference represents quite a large effect, and the confidence limits for Δ are wide.

2.6 Multiple regression

Extension of the linear logistic regression model to multiple regression embodies in principle no new ideas. The main problem in practice is usually the choice of appropriate explanatory variables to include in the fitted model, the general strategy for which is discussed in Appendix 2 and which here is illustrated by some examples.

Example 2.10 Analysis of ingot data (continued)
We extend the analysis of Example 2.2 where a simple logistic model was fitted to heating time, x_1, ignoring the effect of soaking time, x_2. This gave

$$\hat{\lambda} = -5.415 + 0.0807x_1.$$

If we now include x_2 and refit the model, then

$$\hat{\lambda} = -5.559 + 0.0820x_1 + 0.0568x_2$$

with a small but not unexpected change in the coefficient of x_1. The coefficient of x_2, which has a standard error 0.0331, is not significant and implies that the earlier analysis was satisfactory.

The observed proportions of ingots not ready for rolling at the four levels of x_1 are $0/55 = 0.0$, $2/157 = 0.013$, $7/159 = 0.044$, $3/16 = 0.188$. The fitted proportions based on the above model, involving only x_1, are 0.008, 0.014, 0.038, 0.214, agreeing fairly well with the data.

If we transform x_1 and fit instead a logistic model on $\log x_1$ the

See pp. 11, 36, 39-41

i.e., $\theta_i = \dfrac{e^{\alpha + \beta \log x_{1i}}}{1 + e^{\alpha + \beta \log x_{1i}}}$

(cf. p. 37.)

fitted proportions are 0.002, 0.010, 0.047, 0.178, which agree even more closely with the observed pattern of proportions. The maximized log likelihood is 0.38 higher for the regression on $\log x_1$ than for that on x_1, i.e. the likelihood is increased by the modest factor of 1.46. A formal test based solely on this is not straightforward because one model is not nested within the other.

Example 2.11 Formation of crystals in urine
Andrews and Herzberg (1985, p. 249) give data on six physical characteristics relating to 79 specimens of urine for which the absence or presence of calcium oxalate crystals was observed. The characteristics are z_1, specific gravity; z_2, pH; z_3, osmolarity; z_4, conductivity; z_5, urea concentration; z_6, calcium concentration. We write $Y = 0$ or 1 to denote absence or presence of crystals. The data are complete for 77 of the specimens, which we use for analysis.

There is considerable difference in the magnitude of the six variables $z_1, \ldots, z_6$, with some noticeably high values in a number of cases, and high correlations between the variables. The correlation coefficients for (z_1, z_3), (z_1, z_5), (z_3, z_4) and (z_3, z_5) all exceed 0.8.

To improve numerical stability and in the hope of improving linearity we transform the variables, taking $x_1 = 100 \log z_1, x_2 = z_2/10$ (pH being a logarithmic measure), $x_3 = \log z_3$, $x_4 = \log z_4$, $x_5 = \log z_5, x_6 = \log z_6$. This has little effect on the magnitudes of the correlations between the variables.

A fitted model containing all six variables $x_1, \ldots, x_6$ indicates specimen 54 to be an outlier. It has an observed value $Y = 1$ but a fitted probability of only 0.022. Omitting this observation changes the numerical values of the estimated coefficients in the fitted multiple regression, but does not substantially affect their qualitative interpretation or statistical significance. The fitted model based on the remaining 76 specimens is

$$\hat{\lambda} = 25.56 + 4.87x_1 - 1.02x_2 + 1.03x_3 - 5.03x_4 - 4.22x_5 + 2.66x_6.$$

The asymptotic standard errors for the six coefficients are, respectively, 24.3, 2.2, 0.63, 7.0, 3.4, 2.6, 0.7. Augmenting the model by including cross-product terms does not improve the fit of the model. Adding a quadratic term x_5^2 leads to an apparent improvement as measured by the likelihood ratio statistic ($\chi^2 = 4.67$ with 1 degree of freedom) but examination of residuals suggests this is not of practical significance (see Example 2.14).

It does not seem worthwhile omitting the non-significant terms from the above fitted model, given that there are only six variables and a fairly high correlation between them, unless the fitted equation is to be used for prediction purposes and substantial practical saving is involved by measuring fewer than all six physical characteristics. It would be unwise to attempt any detailed interpretation from the fitted model, at least without more background knowledge.

Example 2.12 Clinical trial on patients with cancer of the throat
Kalbfleisch and Prentice (1980, p. 255) give survival times for 195 patients suffering from squamous carcinoma of the oropharynx. Patients at six institutions were randomly allocated to two treatments, radiotherapy alone or radiotherapy plus chemotherapy. There are nine explanatory variables x_1, institution $(1, \ldots, 6)$; x_2, sex; x_3, treatment; x_4, grade of tumour (well differentiated, moderately differentiated, poorly differentiated); x_5, age; x_6, condition (1, no disability; 2, restricted work; 3, requires assistance with self-care; 4, bed confined); x_7, site (faucial arch, tonsillor fossa, pharyngeal tongue); x_8, tumour stage (1, 2 cm or less; 2, 2–4 cm; 3, more than 4 cm; 4, massive invasive tumour); x_9, node stage (no metastases, single node 3 cm or less, single node more than 3 cm, multiple nodes). Very few cases with $x_6 = 3$ or 4 were observed so that we group $x_6 = 2, 3, 4$ together, forming two categories for x_6.

Instead of analysing survival time we consider the probability of survival beyond a specific time t_0. Some of the observed survival times are censored, due mainly to patients surviving throughout the period of the trial. We omit therefore the last 16 patients admitted to the trial, plus 4 others with short censored survival times. This leaves 175 patients for analysis. We take $t_0 = 477$ days, the median survival time for these 175 patients, and write $Y = 1$ if survival time exceeds 477 days, otherwise $Y = 0$.

A major difference between this and Example 2.11 is the nature of the explanatory variables. Here only one of the variables x_5, measuring age, is a straightforward quantitative measurement. Even though x_8 is partially quantitative we treat it and all variables apart from x_5 as categorical, representing them by indicator variables (Appendix A2.3).

The full model involves fitting 20 parameters. The only variables to be statistically significant are x_6, condition (1 parameter) and x_8, tumour stage (3 parameters). A model with only these variables included gives estimated coefficients for x_8 equal to 1.25, 0.41, -0.34, taking the indicator variables to be contrasts with the smallest tumour stage ($x_8 = 1$). A decreasing trend across the categories would have been more meaningful. Few cases are recorded of $x_8 = 1$ and it is sensible to combine categories $x_8 = 1, 2$. Although this leads to a decreasing pattern for a model when $t_0 = 477$ days, it is not so if we consider instead $t_0 = 254$ days, the lower quartile survival time.

If x_8 is dichotomized, combining $x_8 = 1, 2, 3$ into a single category, the fitted model when $t_0 = 477$ days is

$$\hat{\lambda} = 1.64 - 2.28x_6 - 0.879x_8$$

with standard errors for the coefficients, of 0.52, 0.48, 0.35 respectively. In this equation $x_6 = 0, 1$ represent no disability, or some disability, and $x_8 = 0, 1$ represent single primary tumour, or massive invasive tumour. When $t_0 = 254$ days the corresponding fitted model is

$$\hat{\lambda} = 3.56 - 1.83x_6 - 1.34x_8$$

with standard errors 0.66, 0.41, 0.40, respectively. Both models indicate a decreased probability with worsening condition and tumour stage. The standard errors of the estimated coefficients are, somewhat surprisingly, similar in both equations.

Further analysis, e.g. separate analysis for each site, x_7, might be worthwhile. Whether or not the same grouping of the categorical variables, e.g. dichotomizing x_6 or x_8 as above, would be suitable for each analysis is unclear.

2.7 Residuals and diagnostics

2.7.1 Preliminaries

As with other probabilistic models used in the analysis of data, the linear logistic models studied in this monograph are provisional working bases for the analysis rather than rigid specifications to be accepted uncritically. Therefore methods for examining adequacy of fit are important. With relatively large amounts of data graphical analyses and inspection of tables of summary statistics will often lead to a searching examination of the adequacy of a proposed model. With smaller amounts of data, formal tests of significance become relatively more important, primarily because the possibility of apparently appreciable, but nevertheless spurious, departures from the model becomes more important.

In considering these issues for binary data there is often a fairly clear distinction between methods for data on an individual basis and methods for data with substantial grouping of individuals and in which the grouping can be taken as firmly based. Thus looking for a possibly outlying group is not radically different from such analysis with continuous data and relatively minor modifications of the extensive pool of methods for continuous data will be useful. For individual data, however, anomalous observations can be found only in regions where the probability of success is either very high or very low and associated methods for continuous data are not likely to be effective.

In examining the adequacy of models we may look for evidence of

1. failure of the systematic part of the model;
2. inadequacy of the standard simple assumptions about error structure;
3. the existence of outlying anomalous observations;
4. the occurrence of influential observations, i.e. observations whose inclusion or exclusion has a major effect on the primary conclusions of the analysis.

The first of these objectives is the most important.

Note that the recognition that the conclusions from an analysis depend strongly on the response from a particular individual is a reason for checking on the correctness of that response, but is not to be regarded as a reason for automatically 'rejecting' that observation.

There are broadly three ways of checking the adequacy of models:

1. via a single overall test statistic of goodness of fit;
2. via procedures aimed at specific alternatives, including the fitting of extended or modified models;
3. via the use of residuals or other similar statistics designed to detect anomalous or influential observations or unexpected patterns.

In using the last two there will often be some choice between inspection of tables of values, inspection of corresponding graphs and the use of relatively formal test statistics.

All three of these approaches have a role. The disadvantage of the first is that it will not usually give constructive guidance on how to deal with any failure of the original model and is likely to be insensitive in detecting specific types of departure. The advantage of the second method is clear if there is a fairly specific idea of an alternative model, whereas the third has the potential for uncovering the unexpected and the closely associated risks of overinterpretation, i.e. of leading to the pursuit of anomalies that are either accidental features of the data or, even if 'real', are of little or no subject-matter interest.

In some contexts the most natural formal way of checking the correctness of the systematic structure will be clear. Thus in examining linearity of the linear logistic regression $\lambda_i = \alpha + \beta x_i$ a natural approach is to add a quadratic term γx_i^2 and to look at the likelihood ratio test for the hypothesis $\gamma = 0$, or alternatively to compare the maximum likelihood estimate $\hat{\gamma}$ with its standard error. Note, however, that if clear evidence of curvature is found the final conclusions may be best presented in terms of regression on some simple non-linear function, e.g. $\log x$, rather than as a polynomial. Similar ideas apply to multiple regression.

For analyses based on the empirical logistic transform, methods of examining goodness of fit used in normal-theory problems are applicable with minor modifications.

2.7.2 Composite tests

Composite or so-called omnibus tests have to be used with considerable caution, partly, but not primarily, because of their low sensitivity to specific systematic alternatives. If the data are on an individual basis and are not to be grouped, a composite test can be formed by fitting by maximum likelihood the model under test and a larger model in which the number of adjustable parameters is not more than about a fifth of the smaller of the total number of successes and the total number of failures; this last is a rough rule of thumb.

Note that the likelihood ratio test against a saturated alternative in which each individual has a separate parameter is nugatory, because the resulting statistic is a function only of the fitted values under the model.

An alternative to such an overall test is to add the additional parameters one, or a small number, at a time making an allowance for selection in a final judgement of significance. Thus with p quantitative explanatory variables $x_1, \ldots, x_p$, goodness of fit could be tested via the full quadratic model with $\frac{1}{2}p(p+1)$ additional parameters, or by fitting one at a time terms in $x_1^2, \ldots, x_p^2, x_1x_2, \ldots, x_{p-1}x_p$ or some selection thereof.

With data that are either initially grouped, or which are grouped for the purpose of testing goodness of fit, the direct use of Pearson chi-squared or likelihood ratio goodness of fit statistics is available. That is, if the data are in the form of R_i success in m_i trials ($i = 1, \ldots, g$) and the fitted probabilities are $\phi_i(\hat{\beta})$, where β is $p \times 1$ vector of parameters estimated by maximum likelihood, we may use one of

$$X^2 = \sum \frac{\{R_i - m_i\phi_i(\hat{\beta})\}^2}{m_i\phi_i(\hat{\beta})\{1 - \phi_i(\hat{\beta})\}}, \tag{2.80}$$

$$w = 2[\sum R_i \log R_i + \sum (m_i - R_i) \log (m_i - R_i) - \sum m_i \log m_i$$
$$- \sum R_i \log \phi_i(\hat{\beta}) - \sum (m_i - R_i) \log \{1 - \phi_i(\hat{\beta})\}] \tag{2.81}$$

as having asymptotically a chi-squared distribution with $g - p$ degrees of freedom. The second statistic is the likelihood ratio statistic for testing the hypothesized model against the alternative that the ϕ_i are arbitrary, and the asymptotics require that the cell frequencies are not too small.

The relation between the two statistics is best seen by writing

$$R_i = m_i\phi_i(\hat{\beta}) + U_i[m_i\phi_i(\hat{\beta})\{1 - \phi_i(\hat{\beta})\}]^{1/2},$$

so that U_i is a form of standardized deviation. If now the m_i are large and all other quantities are fixed

$$X^2 = \sum U_i^2,$$

$$w = \sum U_i^2 - \frac{1}{6}\sum \frac{\{1 - 2\phi_i(\hat{\beta})\}}{[\phi_i(\hat{\beta})\{1 - \phi_i(\hat{\beta})\}]^{1/2}} \frac{U_i^3}{\sqrt{m_i}} + O\left(\frac{1}{m_i}\right),$$

showing that at least when the m_i are not too small, and especially when the fitted probabilities are close to $\frac{1}{2}$, the statistics are unlikely to differ materially. McCullagh (1986) has given corrections to improve the distributional approximation by chi-squared. Various numerical studies have shown no clear basis for preference between the two statistics.

2.7.3 Residuals

As noted in Section 2.7.1, if the data are substantially grouped it will be necessary only to define a residual for each group. This can be done either via a direct comparison of the number of successes in each group with the number to be expected under the fitted model or via a comparison of the corresponding empirical logistic transform, i.e. via

$$(R_i - m_i\hat{\phi}_i)/\{m_i\hat{\phi}_i(1 - \hat{\phi}_i)\}^{1/2} \tag{2.82}$$

or via the corresponding definition using the empirical logistic transform, namely via (2.26) and (2.27)

$$[Z_i - \log\{\hat{\phi}_i/(1 - \hat{\phi}_i)\}]/\sqrt{V_i}, \tag{2.83}$$

where $\hat{\phi}_i$ denotes the fitted value under the model being assessed. These can be plotted against new or original explanatory variables, against fitted values or in a normal probability plot.

If we are working initially with individual binary observations the definition (2.82) can still be applied now in the form

$$D_i = (Y_i - \hat{\theta}_i)/\{\hat{\theta}_i(1 - \hat{\theta}_i)\}^{1/2}. \tag{2.84}$$

The individual D_i are, however, virtually useless, because at each value of $\hat{\theta}_i$ only two possible values of the residual can arise. It is therefore essential for interpretation that the residuals are grouped, typically by adding together in rational sets, based for example on adjacency in the space of explanatory variables; when t such residuals are added it is sensible to divide by $\sqrt{t}$ to preserve the unit variance and the interpretation via an approximate standard normal distribution.

Example 2.13 Trend in a 2×2 contingency table (continued)
The main application of graphical methods using residuals is likely to be in the reduction of extensive data. Here we give a simple example based on the data of Table 2.6, continuing the discussion of Example 2.9. Residuals can be calculated after the fitting by maximum likelihood of any of the models of Example 2.9. Here we consider model IV, in which the logistic transform on the ith trial is

$$\lambda_i = \begin{cases} \mu - \delta + \beta(i - 15) & \text{for treatment 0,} \\ \mu + \delta + \beta(i - 15) & \text{for treatment 1.} \end{cases}$$

From the maximum likelihood estimates $\hat{\mu}$, $\hat{\beta}$ and $\hat{\delta}$, the fitted transforms, $\hat{\lambda}_i$, can be calculated, leading to the corresponding fitted probabilities, $\hat{\theta}_i$. The values, $\hat{\theta}_i$, and the observed residuals d_i, defined by (2.84), are given in the last two columns of Table 2.6.

The fitted values are of interest in themselves as one clear way of indicating the meaning of the fitted model. The residuals can be plotted in various ways, for example against the value of any other regressor variable that might be relevant. In Fig. 2.1(a) the residuals are plotted against serial order, values corresponding to the two treatments being distinguished.

The absence of trends indicates the adequacy of the model, although such diagrams have to be interpreted cautiously. Because of the binary nature of the responses, the residuals, although scaled to

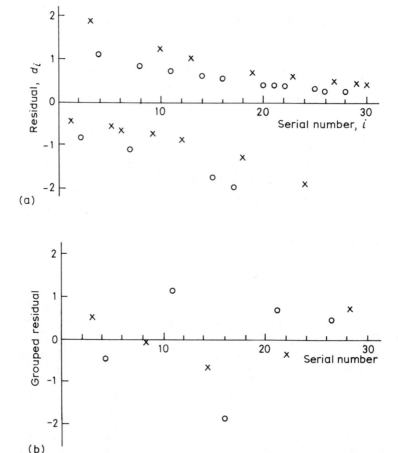

Figure 2.1 *Two treatments with trend. Plots of residuals: (a) individual residuals, d_i:* $\bigcirc = $ *treatment* 0, $\times = $ *treatment* 1; *(b) grouped residuals:* $\bigcirc = $ *treatment* 0; $\times = $ *treatment* 1.

have zero mean and unit standard deviation, have a very non-normal distribution. In particular, residuals very close to zero do not occur, except for extreme values of $\hat{\theta}_i$. In such regions of high or low probability of success, the residuals have a very skew distribution. Thus if the probability of success is high, the residuals are either small and positive or large and negative. The main feature of Fig. 2.1(a) perhaps calling for comment is the pair of low values on trials 15 and 17, corresponding to two failures under treatment 0 in a region of fairly high probability of success.

The inevitable systematic effects in diagrams such as Fig. 2.1(a) can be obviated to some extent by combining residuals in comparable blocks. In Fig. 2.1(b) grouped residuals are shown. They are defined as $\sqrt{3}$ times the average of adjacent sets of three residuals, for the same treatment. Thus the first three residuals for treatment 0 are -0.80, 1.11 and -1.08, and the grouped residuals give $\sqrt{3} \times (-0.77/3) = -0.44$. The factor $\sqrt{3}$ is included to make the variance approximately one.

We shall not discuss in detail the calculation of test statistics from the residuals. If we wished to test whether the regressions on serial order have the same logistic slope for the two treatments, we could calculate a regression coefficient for residuals on serial order separately for the two treatments, and form a test statistic from their difference. In the present case this would clearly confirm the conclusion of Example 2.9, that there is no evidence from these data that the introduction of separate parameters is required.

Example 2.14 Formation of crystals in urine (continued)
In Example 2.11 we fitted a logistic model which was linear in six explanatory variables and the addition of a quadratic term x_5^2 appeared to be significant as measured by the likelihood ratio statistic. Figure 2.2 plots grouped residuals against x_5. The residuals are calculated from the linear model in the six variables. They are defined initially by (2.84) and then grouped, taking $\sqrt{5}$ times the average of adjacent sets of five residuals. The noticeably high residual with value 3.28, and the preceding plotted value, contain three neighbouring points for which the observed responses are $y = 1$ but the predicted probabilities are small, and it is these points which contribute to the apparent significance of x_5^2 showing the danger of misinterpreting an anomaly discovered via the quadratic term.

Example 2.14 confirms that allowance for selection needs to be

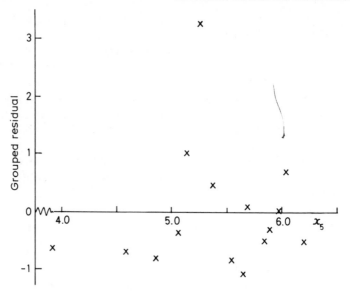

Figure 2.2 *Formation of crystals in urine. Grouped residuals plotted against x_5.*

made when checking the adequacy of the linear model by fitting additional terms selected one at a time from the full quadratic model. With explanatory variables which are so highly correlated it would be difficult to determine what allowance to make, but examination of residuals, as in Fig. 2.2, provides a simple means of checking for spurious results.

Example 2.15 Association between smoking and lung cancer
Cornfield (1956) reproduced and analysed the data in Table 2.8, given originally by Dorn (1954). Cornfield's analysis was based directly on the likelihood of the observations as a function of the differences $\Delta_1, \ldots, \Delta_k$; here we illustrate the use of the empirical logistic transform.

Table 2.9 gives estimates, $\tilde{\Delta}_s^{(w)}$, of the logistic difference for each study based on the weighted form of the empirical logistic transform, (2.28). Also the table shows the variance, (2.71), the standardized residual from the overall weighted mean and, for a purpose that will appear later, the sum of the logistic transforms for lung cancer and for control groups. The argument of Section 4.3 shows that the $\tilde{\Delta}_s^{(w)}$ can

Table 2.8 *Fourteen retrospective studies on the association between smoking and lung cancer*

Study	Lung cancer patients		Control patients	
	Total	Non-smokers	Total	Non-smokers
1	86	3	86	14
2	93	3	270	43
3	136	7	100	19
4	82	12	522	125
5	444	32	430	131
6	605	8	780	114
7	93	5	186	12
8	1357	7	1357	61
9	63	3	133	27
10	477	18	615	81
11	728	4	300	54
12	518	19	518	56
13	490	39	2365	636
14	265	5	287	28

Table 2.9 *Logistic analysis of the data of Table 2.8*

Study, s	$\tilde{\Delta}_s^{(w)}$	$\tilde{v}_s^{(w)}$	Residual	Logistic sum
1	1.83	0.426	0.37	-5.16
2	1.90	0.368	0.53	-5.25
3	1.51	0.214	-0.16	-4.45
4	0.64	0.107	-2.89	-2.96
5	1.74	0.045	0.74	-3.40
6	2.61	0.137	2.76	-6.14
7	0.25	0.298	-2.44	-5.68
8	2.27	0.161	1.71	-8.40
9	1.79	0.391	0.32	-4.55
10	1.37	0.072	-0.79	-5.16
11	3.81	0.274	4.25	-6.85
12	1.18	0.075	-1.50	-5.41
13	1.46	0.030	-0.73	-3.46
14	1.81	0.243	0.46	-6.30
Weighted mean	1.585			

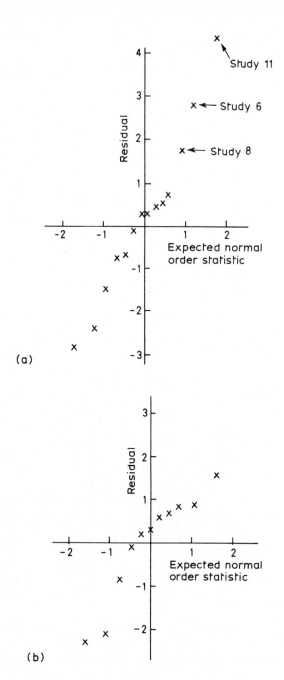

(a)

(b)

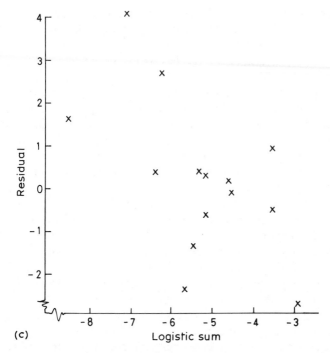

(c)

Figure 2.3 *Graphical analyses of data of Table 2.8: (a) normal plot of residuals; (b) normal plot of residuals omitting studies 6, 8 and 11; (c) residual plotted against logistic sum.*

be regarded as estimating also the logistic difference in the proportions not suffering from lung cancer as between two populations, one of non-smokers and one of smokers.

The data provide strong evidence against the constancy of the logistic difference. Some individual residuals are much too large, although that for study 11 is suspect because the number of non-smokers in the lung cancer group is only 4. The residual sum of squares is 47.7 to be compared approximately with chi-squared with 13 degrees of freedom.

Although a full interpretation would require detailed knowledge of the individual studies, some of the techniques for a further analysis can be illustrated.

The ranked logistic residuals of Table 2.9 are plotted against the expected normal order statistics in Fig. 2.3(a). Three values stand out, corresponding to studies 8, 6 and 11. When these are omitted, and residuals recalculated from the mean of the remaining eleven studies, Fig. 2.3(b) is obtained. The resulting residual sum of squares is 15.1 with 10 degrees of freedom, and, while there are two rather large negative residuals, much of the systematic variation has been removed. Unit slope is expected.

As noted in Section 2.5, an easily implemented likelihood ratio test of homogeneity can be based on the saddle-point approximation to the conditional log likelihood function. This gives values of 55.2 with 13 degrees of freedom and 13.6 with 10 degrees of freedom, as compared with the above values of 47.7 and 15.1; any major difference between the two approaches is likely to emerge only when smaller frequencies are involved.

For further interpretation the estimated $\tilde{\Delta}_s^{(w)}$'s can be plotted against variables characterizing the individual studies. The only such variable available here is the overall level of response in the study, conveniently measured by the sum of the logistic transforms in lung cancer and control groups and given in the last column of Table 2.9. Figure 2.3(c) shows the result of plotting residual based on the difference of logistic transforms against the sums. Note that because the individual transforms have unequal variances, the sums and differences are correlated. Since, however, the variance in the sums is large compared with the individual sampling errors this correlation has no influence on the qualitative interpretation of Fig. 2.3(c).

There is some suggestion from Fig. 2.3(c) that the three studies, 8, 6 and 11, have not only large differences between control and lung cancer groups but also have large negative logistic sums, i.e. overall low proportions of non-smokers. Thus the largest logistic differences between control and lung cancer groups have been found in studies in which the overall proportion of non-smokers is low. Taken with the result of Section 4.3 this suggests that the largest proportional effect of smoking on lung cancer may be found in contexts in which the overall proportion of smokers is high. Of course, this is no more than a tentative suggestion obtained from an informal graphical analysis and the explanation of the differences between studies may well lie in some other features of the populations under study.

We have used here the simplest and most direct definition of residuals. A number of more complicated definitions are available essentially designed to improve the distributional properties of the individual residuals. Two different ideas which can be used separately or together are

(a) to base the definition of residual more directly on log likelihood via the so-called deviance residual

$$D_i^{(l)} = \sqrt{2} \operatorname{sgn}(R_i - m_i\hat{\phi}_i)\{l_i(\phi_i = R_i/m_i) - l_i(\hat{\phi}_i)\}^{1/2}, \quad (2.85)$$

comparing the contribution of the ith observation or group to the log likelihood at R_i/m_i, the maximum likelihood of the group probability based solely on y_i with that based on $\hat{\phi}_i$ under the fitted model;

(b) to standardize D_i or $D_i^{(l)}$ by dividing by $\sqrt{(1 - h_i)}$, where h_i is the ith diagonal element of the matrix

$$H = W^{1/2}X(X^T W X)^{-1}X^T W^{1/2}, \quad (2.86)$$

and $W = \operatorname{diag}\{m_i\hat{\phi}_i(1 - \hat{\phi}_i)\}$.

The standardization approximately corrects the residual to have unit variance, allowing for errors of estimation in $\hat{\phi}_i$.

2.7.4 Influence diagnostics

For normal-theory linear models there is an exact explicit relation between the least squares estimates from fitting a model to n observations and the result of fitting the same model to $(n-1)$ observations, a specific observation having been deleted. This helps the identification of influential observations, i.e. ones whose deletion induces a major change. Note, however, that there can be complicated masking effects if several observations are influential.

For logistic and similar models there is no corresponding exact relation. Pregibon (1981) has, however, used the relation between the methods of maximum likelihood and of weighted least squares to show that in the linear logistic model $\lambda = x\beta$, the effect of deleting the ith observation is approximately to change the maximum likelihood estimate $\hat{\beta}$ to

$$\hat{\beta}_{(i)} \simeq \hat{\beta} - w_i^{1/2}(1 - h_i)^{-1}d_i(X^T W X)^{-1}x_i. \quad (2.87)$$

From this one can produce a measure of the change of log likelihood or equivalently the displacement of the confidence contours for β, namely

$$c_i = \frac{d_i^2 h_i}{(p+1)(1-h_i)^2}, \quad (2.88)$$

where $p + 1$ is the number of parameters fitted. Large values of c_i identify the most influential observations. These should on no account be automatically discarded, but it will often be a sensible precaution to see whether their omission in fact has a serious effect. Usually only observations with $c_i > 1$ need be considered as candidates for omission.

Yet another variant of these ideas is that of a partial residual plot, this being designed not merely to detect a relation with a new explanatory variable, or anomalous relation with an old explanatory variables, but to show more directly than Figs 2.2 or 2.3 the form of the correct relation. See the Bibliographic Notes and Exercise 39 for details.

Example 2.16 Fertilizer experiment on growth of cauliflowers
(Cox and Snell, 1981, p. 112)
In an experiment on the effect of nitrogen and potassium upon the growth of cauliflowers, four levels of nitrogen and two levels of potassium were tested:
   Nitrogen levels: 0, 60, 120, 180 units per acre (coded as 0, 1, 2, 3);
   Potassium levels: 200, 300 units per acre (coded as A, B).

The experiment was arranged in 4 blocks, each containing 4 plots as shown below. When harvested, the cauliflowers were graded according to size. Table 2.10 shows the yield (number of cauliflowers) of different sizes: grade 12, for example, means that 12 cauliflowers fit into a standard size crate.

We take as response variable the proportion of cauliflowers of grade 24 or better, i.e.

$$(n_{12} + n_{16} + n_{24})/48,$$

where n_r ($r = 12, 16, 24, 30$) denotes the observed frequency of cauliflowers of grade r. The divisor 48 corresponds to the total number of cauliflowers per plot, except in those instances in which some plants died.

The experimental design is a complete 2×4 factorial, confounded between blocks I and II, with blocks III and IV as a replicate. If we write $K = -1, 1$ to represent the two levels of potassium and $N_Q = -1, 1, 1, -1$ to represent a quadratic component across the four levels of nitrogen, then the interaction $K \times N_Q$ defines the confounding.

We can fit a multiple logistic regression model involving the

Table 2.10 *Numbers of cauliflowers of each grade*

Block	Treatment	Grade				Unmarketable
		12	16	24	30	
I	0A	—	1	21	24	2
	2B	1	6	24	13	4
	1B	—	4	28	12	4
	3A	1	10	26	9	1
II	3B	—	4	26	14	4
	1A	—	5	27	13	3
	0B	—	—	12	28	8
	2A	—	5	35	5	3
III	1B	—	1	22	22	3
	0A	—	1	8	33	3
	3A	—	6	22	17	2
	2B	—	3	27	14	4
IV	0B	—	—	8	30	10
	2A	—	7	16	22	3
	3B	—	2	31	11	4
	1A	—	—	13	26	9

contrasts of prime interest K, N_L, $K \times N_L$ and N_Q, plus $K \times N_Q$, R and $R \times K \times N_Q$, these last three accounting for the three degrees of freedom between blocks. Here N_L denotes a linear contrast for nitrogen and R a contrast between replicates. The fitted model has a likelihood ratio statistic against a saturated model of $\chi^2 = 20.61$, with 8 degrees of freedom, which suggests some lack of fit in the model. Use of the likelihood statistic is justified since the proportions under analysis are not too close to 0 or 1.

The influence diagnostic (2.88) is greatest for line 14 of the data, for which $c = 2.78$. Omitting this line and refitting the model reduces the likelihood statistic to 8.94, with 7 degrees of freedom. However, this changes the coefficient of the potassium contrast, which now is of borderline significance, but is in a negative direction and is unlikely to be meaningful.

Even with appreciable background knowledge it may not be easy to decide what to do about outliers or influential observations. In the present case, in the absence of background knowledge, it seems reasonably cautious to base our conclusions on an analysis of all sixteen observations, accepting that the variability in the data is greater than is assumed by a binomial model (see Section 3.2).

2.8 Factorial arrangements

In Section 2.6 we studied the dependence of a binary response on a set of explanatory variables via techniques analogous to those of multiple linear regression. We now suppose that the explanatory variables have factorial structure, i.e. are formed via a number of factors each at a relatively small number of levels, all or most of the combinations of factor levels being observable.

The special simplifications that arise in the analysis of balanced factorial systems with quantitative responses will not often apply to binary data, except as a rather rough approximation. For, even if there are equal numbers of individuals in each cell, unequal variances will be encountered, except in the rather uninteresting case where the system under study is close to a totally null one without systematic differences.

Nevertheless many of the general considerations applying to factorial arrangements continue to hold. In particular

1. Factors can be regarded as treatments (or in observational studies as quasi-treatments), as intrinsic factors describing the individuals under study or the environment encountered or as non-specific factors defining strata or other groupings of the material.

2. If the factor levels, especially of treatment factors, are defined by a quantitative variable or are ordinal, this needs to be taken into account in interpretation, at least informally.

3. Interest will usually focus on the treatment factors and their interaction with intrinsic factors, and often in finding a representation in which there are few if any interactions involving treatment factors.

There are of course exceptions to (3), notably when one or two cells of the factorial structure show strongly anomalous response.

Table 2.11 *2×2 Factorial arrangement with binary response. Columns 2 and 3, numbers of successes and numbers of trials. Columns 4 and 5, corresponding probabilities*

		Intrinsic factor (sex)			
		0	1	0	1
Treatment	0	r_{00}/n_{00}	r_{01}/n_{01}	θ_{00}	θ_{01}
	1	r_{10}/n_{10}	r_{11}/n_{11}	θ_{10}	θ_{11}

As the simplest non-trivial example, suppose that there are two two-level factors, one treatment factor (0 and 1, control and treatment, say) and one intrinsic factor, for example sex (0 and 1, male and female). Thus with a binary response we can set the results and the corresponding probabilities out as in Table 2.11.

Incidentally note that if we abandoned the distinction between response and explanatory variables, the data would be regarded as a $2 \times 2 \times 2$ table. Not only shall we not do this in the present context but further the distinction between treatment and intrinsic factors suggests that it would normally be best to concentrate on study of whatever treatment effect is present. If the treatment effect is appreciably different for men and for women (treatment $\times$ sex interaction) then the conclusions will be summarized via separate treatment effects for men and for women, whereas if the treatment effect is reasonably constant a single main effect of treatment can be calculated. In this situation the main effect of sex would often be of little interest.

Sometimes careful inspection of tables of proportions of successes will provide an adequate analysis, but if we need to proceed more formally mathematical representation of the probabilities in Table 2.11 is needed. This may be done via θ, or $\Phi^{-1}(\theta), \ldots,$ although, in line with the main emphasis of the book we shall concentrate on $\lambda = \log \{\theta/(1-\theta)\}$; for the analysis of a single such set of data it would rarely matter much which version is used, but in more complicated cases a scale in which the treatment effect is constant would normally be preferred.

In analysing formally Table 2.11 we would normally fit by maximum likelihood the models

1. arbitrary θ_{sr}, four parameters, thus allowing for treatment $\times$ sex interaction;
2. the model with treatment and sex main effects on the logistic (or other) scale, namely

$$\lambda_{00} = \mu - \alpha - \beta, \qquad \lambda_{01} = \mu - \alpha + \beta,$$
$$\lambda_{10} = \mu + \alpha - \beta, \qquad \lambda_{11} = \mu + \alpha + \beta; \qquad (2.89)$$

   with three parameters;
3. the model with only sex effect, two parameters.

Model (1) is equivalent to separate analyses of males and females and the most meaningful expression of the conclusions from such a model is via separate estimated treatment effects for males and females. Comparison of maximized log likelihoods allows the testing of null hypotheses of absence of treatment × sex interaction and of absence of treatment main effect (assuming absence of interaction). Note that it would very rarely be meaningful to test the null hypothesis of zero treatment main effect in the presence of treatment × sex interaction; also it would be possible to test the null hypothesis of no sex difference assuming no treatment × sex interaction or assuming no treatment effects at all, but this would often not be a primary object of the analysis.

Usually, even if the data are consistent with total absence of treatment effects, it will be sensible to give approximate confidence limits for a treatment main effect. Moreover, in summarizing conclusions for possible re-analysis at a later stage in conjunction with possible further relevant data, it will often be wise to retain proportions of successes in the two treatment groups cross-classified with each of the more important intrinsic factors in turn, even if the corresponding two-factor interactions are far from significant.

The models just described can all be fitted by various standard library packages; see Section A3.2. Note, however, that in many cases the output will not be in a form for incisive interpretation along the lines just outlined and simple hand calculation may be needed as a supplement.

Very similar methods based on the above principles apply to general factorial arrangements. The main problem in higher-dimensional problems is that, the likelihood ratio statistic for testing the hypothesis that say, a particular interaction is absent depends on the choice of parameters in the model, as is the case, of course, in the study via analysis of variance of unbalanced factorial arrangements of normally distributed responses.

When a saturated model is fitted maximum likelihood estimation is in effect by equating observed proportions and theoretical probabilities. For logistic models this means the calculation of a full set of standard factorial contrasts from the empirical logistic transforms (assuming the absence of zero or 100% responses).

Example 2.17 A 2 × (3 × 2²) factorial experiment
Table 2.12 gives the data from $2 \times (3 \times 2^2)$ experiment comparing two detergents, a new product X and a standard product M (Ries and Smith, 1963). The three factors are water softness, at three levels, temperature, at two levels, and a factor whose two levels correspond

Table 2.12 *Number r_j of prefrences for brand X our of n_j individuals*

Water softness		M previous non-user Temperature		M previous user Temperature	
		Low	High	Low	High
Hard	r_j	68	42	37	24
	n_j	110	72	89	67
	z_j	0.477	0.332	−0.337	−0.575
	v_j	0.0381	0.0563	0.0457	0.0637
Medium	r_j	66	33	47	23
	n_j	116	56	102	70
	z_j	0.275	0.355	−0.156	−0.704
	v_j	0.0348	0.0723	0.0391	0.0634
Soft	r_j	63	29	57	19
	n_j	116	56	106	48
	z_j	0.171	0.070	0.150	−0.414
	v_j	0.0344	0.0703	0.0376	0.0851

to previous experience and no previous experience with M. For each of the twelve factor combinations, a number, n_j, of individuals, between 48 and 116, use both detergents and r_j of these prefer X, the remainder preferring M.

To construct a set of single degrees of freedom, we suppose that the three levels of water softness are in some sense equally spaced and so split all contrasts involving softness into linear and quadratic components. The quantities z_j and v_j, computed by (2.26) and (2.27), are given in Table 2.12. Next, standardized contrasts are computed. For example, for the quadratic main effect of softness, we compute first

$$(0.477 + 0.332 - 0.337 - 0.575)$$
$$- 2(0.275 + 0.355 - 0.156 - 0.704) \qquad (2.90)$$
$$+ (0.171 + 0.070 + 0.150 - 0.414) = 0.334.$$

The estimated variance is

$$(0.0381 + \cdots + 0.0637) + 4(0.0348 + \cdots + 0.0634)$$
$$+ (0.0344 + \cdots + 0.0851) = 1.270,$$

and finally the standardized contrast is

$$\frac{0.334}{\sqrt{1.270}} = 0.296;$$

Table 2.13 *Logistic factorial standardized contrasts estimated from the data of table 12.12*

Temperature T	-1.894
User vs. non-user M	-4.642
Softness, linear S_L	-0.122
quadratic S_Q	0.296
$T \times M$	-1.479
$T \times S_L$	0.429
$T \times S_Q$	0.099
$M \times S_L$	-1.852
$M \times S_Q$	0.669
$T \times M \times S_L$	0.563
$T \times M \times S_Q$	0.621

this has approximately unit theoretical variance. The contrasts are collected in Table 2.13.

A preliminary plot, or inspection of Table 2.13, shows that the main effect of M is very highly significant; note that in Table 2.12 z_j for a non-user cell is always greater than z_j for the corresponding user cell. Therefore the main effect of M was omitted from the plot and Fig. 2.4 shows the result; the line corresponds to the theoretical unit variance. There are three suspicious points, namely, in order from the highest: T, $M \times S_L$ and $M \times T$. While these are of borderline significance individually, the fact that three of the large contrasts involve M and that this is a factor corresponding to a classification of the experimental units (individuals) rather than to a treatment, suggests splitting the experiment on factor M, i.e., analysing separately the results from the two levels of M. The good agreement of the remaining points with the theoretical line is some check on the absence of additional components of error.

To do this, we analyse separately the two halves of Table 2.12 corresponding to the two levels of M. Note that a positive z_j corresponds to an average preference for X over M, and a negative z_j to an average preference for M over X. The conclusions are briefly as follows. For previous non-users of M, all cells show an average preference for X, i.e. positive z_j. There is no systematic change with temperature, but a decrease in preference with water softness, the average difference in z_j between soft and hard being -0.284 ± 0.223, i.e. the approximate standard error estimated from the v_j's is 0.223. For previous users of M, all cells except one show an average preference for M, i.e. a negative z_j. The preference is stronger at the

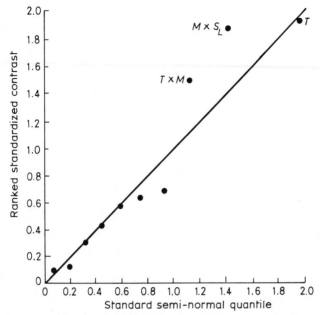

Figure 2.4 *Half-normal plot for logistic factorial contrasts from Table 2.13, omitting main effect of M. Line gives theoretical slope.*

higher temperature, the average difference of z_j between high and low temperatures being -0.450 ± 0.193. There is a general decrease in strength of preference for M with water softness; in terms of z_j this is an effect of opposite sign from that observed with previous non-users of M. The average difference in z_j between soft and hard for previous users of M is 0.324 ± 0.241. Thus in both cases the effects of softness, while suggestive, are not clearly established.

The above are broadly the conclusions reached by Ries and Smith (1963) by a series of chi-squared tests. The present approach, however, allows the estimation as well as the significance testing of effects.

The rather informal graphical approach is likely to be relatively more useful in more complex systems. Then some rough preliminary analysis will usually be necessary before fitting a suitably simplified model either by maximum likelihood (Dyke and Patterson, 1952) or by weighted least squares. For an analysis based directly on the proportions, see Cox and Snell (1981, p. 107).

If it appears that one of the non-treatment factors has little or no

Table 2.14 *Artificial example illustrating Simpson's paradox*

		$X_2 = 0$		$X_2 = 1$	
		Y		Y	
		0	1	0	1
X_1	0	50	950	5000	5000
	1	1000	9000	95	5

effect on response, it will be tempting to analyse the data collapsed over that factor, i.e. to calculate frequencies summed over all levels of that factor and thenceforth to ignore the factor, i.e. to proceed as if that factor had not been observed. While this is often helpful, especially when there are a large number of factors, care is needed in very unbalanced data because collapsing over one factor can reverse the direction of an effect of another factor. This possibility, called Simpson's paradox, is best illustrated by an artificial numerical example; see Table 2.14. Proportions of cases in which $Y = 1$ from Table 2.14 are such that, collapsing over X_2, we have

$$\text{prop}(Y = 1 | X_1 = 0) = 0.46 < \text{prop}(Y = 1 | X_1 = 1) = 0.89,$$

although at each level of X_2 the opposite inequality holds, namely

$$\text{prop}(Y = 1 | X_1 = 0, X_2 = 0) > \text{prop}(Y = 1 | X_1 = 1, X_2 = 0),$$
$$\text{prop}(Y = 1 | X_1 = 0, X_2 = 1) > \text{prop}(Y = 1 | X_1 = 1, X_2 = 1).$$

The explanation lies in the fact that collapsing over X_2 is highly unbalanced with respect to X_1.

Example 2.18 is a further illustration of the analysis of this kind of data.

Example 2.18 Educational plans of Wisconsin schoolboys
Sewell and Shah (1968) have investigated for some Wisconsin high school 'senior' boys and girls the relationship between variables:

1. socio-economic status (high, upper middle, lower middle, low);
2   intelligence (high, upper middle, lower middle, low);
3. parental encouragement (low, high);
4. plans for attending college (yes, no).

The data for boys are given in Table 2.15.

Table 2.15 *Socioeconomic status, intelligence, parental encouragement and college plans for Wisconsin schoolboys*

IQ	College plans	Parental encouragement	SES			
			L	LM	UM	H
L	Yes	Low	4	2	8	4
		High	13	27	47	39
	No	Low	349	232	166	48
		High	64	84	91	57
LM	Yes	Low	9	7	6	5
		High	33	64	74	123
	No	Low	207	201	120	47
		High	72	95	110	90
UM	Yes	Low	12	12	17	9
		High	38	93	148	224
	No	Low	126	115	92	41
		High	54	92	100	65
H	Yes	Low	10	17	6	8
		High	49	119	198	414
	No	Low	67	79	42	17
		High	43	59	73	54

Table 2.16 *Wisconsin schoolboys. Estimated parameters and standard errors for the main effects model*

Parameter		Estimate	Standard error
SES	L	0	—
	LM	0.36	0.12
	UM	0.66	0.12
	H	1.41	0.12
IQ	L	0	—
	LM	0.59	0.12
	UM	1.33	0.12
	H	1.97	0.12
PE	L	0	—
	H	2.46	0.10

GLIM sets the first parameter of each group to zero.

Cox and Snell (1981, p. 162) studied the dependence of CP (college plans) as a response on the other variables as explanatory variables and found the data to be fitted adequately by a linear logistic model containing the three main effects (SES, IQ and PE), with no interactions; the difference in 2 log (maximized likelihood) between this and the full model giving perfect fit is 25.24 with 24 degrees of freedom. Each of the main effects is statistically significant; the omission of any one main effect from the model results in a significant increase in the 2 log (maximized likelihood) statistic.

The estimated parameters and standard errors are given in Table 2.16. There is a marked trend across the estimates for SES, the higher the socioeconomic status the higher the probability of plans to attend college. Increased level of intelligence and parental encouragement likewise increase the probability of college plans.

2.9 Cross-over designs

Designs in which physically the same individual is used as an experimental unit on more than one occasion are called cross-over designs. They are valuable in several fields of application, notably animal nutrition, experimental psychology and clinical trials. By ensuring that the primary comparisons of interest are made within rather than between individuals, substantial gains are in principle possible.

The simplest example of a cross-over trial has two treatments and two periods. That is, the individuals are randomly allocated to one of two groups, the first receiving T_0 in the first period and T_1 in the second, the second group receiving the treatments in the reverse order T_1, T_0. Even with two treatments there are other possibilities, such as allocating some individuals to a constant treatment regime, some to T_0, T_0 and others to T_1, T_1; also if three periods can be used some extra analytical insight can be gained. We shall not consider these possibilities. With more than two treatments many types of design can be considered.

Difficulties with all cross-over designs, in whatever field of application, are connected with the possibility that treatment effects in a regime of rapidly changing treatments may be different from those when a constant treatment is applied to each individual for an extended period, and in particular with the possibility of carry-over or residual effects, i.e. that the effect of a treatment persists beyond the period in which it is applied. To some extent the latter difficulty can be obviated by the insertion of 'wash-out' periods between the experimental treatments, in which individuals are supposedly restored to a baseline state. Even if this is checked by suitable supplementary measurements, some doubt will often remain; direct test of the presence of carry-over effects, or of treatment × period interaction, is possible but usually of low sensitivity.

Occasionally the possible presence of residual effects is of intrinsic interest, but most commonly the objective is the direct comparison of treatments in a context in which it is thought on general grounds that carry-over effects are unlikely and in which it is reasonable to hope for a substantial gain in precision from the design. It is then usually best to adopt the following broad strategy. One analyses primarily in the light of a model with direct treatment and period effects and allowing for systematic difference between individuals. A check for compatability with simple assumptions is carried out and if that yields evidence of interaction then reconsideration of the whole design is desirable. One possibility is then to analyse the first period observations as a design on their own, although precision will normally be low.

We concentrate now on the two-treatment two-period design with binary response. The analysis is an extension of the matched pair analysis discussed in Section 2.4. A direct analogue of the model of Section 2.4 is as follows. For the ith individual the logit probabilities of the responses in periods 1, 2 are respectively in the first group (T_0, T_1)

$$\alpha_i, \qquad \alpha_i + \Delta + \pi,$$

whereas in the second group the logit probabilities are

$$\alpha_i + \Delta, \qquad \alpha_i + \pi,$$

where Δ and π are respectively treatment and period effects.

It follows by the argument of Section 2.4 that only the mixed responses are relevant to inference about Δ and π and thereby we obtain two conditional binomial distributions

Group I: $R_{01}^{(1)}$ given $R_{01}^{(1)} + R_{10}^{(1)}$ has parameter $e^{\Delta + \pi}/(1 + e^{\Delta + \pi})$;
Group II: $R_{01}^{(2)}$ given $R_{01}^{(2)} + R_{10}^{(2)}$ has parameter $e^{-\Delta + \pi}/(1 + e^{-\Delta + \pi})$.

Thus in the 2×2 table

$$R_{01}^{(1)} \quad R_{10}^{(1)} : R_{01}^{(1)} + R_{10}^{(1)}$$
$$R_{01}^{(2)} \quad R_{10}^{(2)} : R_{01}^{(2)} + R_{10}^{(2)}$$

the odds ratio is $e^{2\Delta}$ and simple inference methods for the 2×2 table are applicable, including a test of $\Delta = 0$. If, however, we rewrite the table

$$
\begin{array}{ll}
R_{01}^{(1)} \quad R_{10}^{(1)} \; : \; R_{01}^{(1)} + R_{10}^{(1)} \\
R_{10}^{(2)} \quad R_{01}^{(2)} \; : \; R_{01}^{(2)} + R_{10}^{(2)}
\end{array}
$$

the odds ratio is $e^{2\pi}$ and the period effect can be examined.

From this context there is no more information; the insertion of a large number of nuisance parameters, one for each individual, and the consequent enforced conditioning mean that only the 'mixed' responses can be used and, for example, treatment $\times$ period interaction cannot be tested. Yet if individuals have been randomized to the two groups it is clear that some further information can be recovered. This is probably best approached by an extension of the second set of results in Section 2.4; namely, we regard each individual as determining one of four possible responses and describe the situation initially by six independent probabilities

$$
\begin{array}{llll}
\theta_{00}^{(1)}, & \theta_{01}^{(1)}, & \theta_{10}^{(1)}, & \theta_{11}^{(1)}, & \sum \theta_{ij}^{(1)} = 1; \\
\theta_{00}^{(2)}, & \theta_{01}^{(2)}, & \theta_{10}^{(2)}, & \theta_{11}^{(2)}, & \sum \theta_{ij}^{(2)} = 1.
\end{array}
$$

The relation between the probabilities in the two groups would then be studied via meaningful contrast.

2.10 Rasch model

We now consider briefly the analysis of binary data in a balanced two-way arrangement. Similar ideas apply to higher-dimensional situations. The most important single application is in educational testing where the rows of the data matrix correspond to individuals under test and the columns to test items, the response in row i and column j being 1 for a correct answer and 0 for an incorrect answer. We represent this situation by an $r \times c$ array of random variables (Y_{ij}), where the rows and columns, initially at least, lack any specific structure. In the educational testing situation both r and c will be quite large and r possibly very large.

In line with models studied earlier a natural starting point is to take the Y_{ij} as independent random variables with

$$
\lambda_{ij} = \alpha_i + \beta_j \tag{2.91}
$$

with some constraint, such as $\sum \alpha_i = 0$, applied where necessary. In the educational testing context α_i measures the ability of the ith individual and β_j the easiness of the jth item. In this context (2.91) is called the Rasch model. From an empirical viewpoint this is the simplest model incorporating row and column effects.

One interpretation and special motivation of (2.91) is via an analogue of tolerance distributions, i.e. via a latent continuous structure. Suppose that there is an unobserved continuous one-dimensional scale of ability, each individual being associated with one point on that scale. Now suppose that for an individual of ability α the probability of correct response to item j is

$$e^{\alpha + \beta_j}/(1 + e^{\alpha + \beta_j}). \tag{2.92}$$

Thus the 50% point of response for item j is for individuals of ability $-\beta_j$. That is, (2.92) defines a series of parallel logistic response curves for the different items, the location of each curve depending on the easiness of the item. If the ability of the ith individual is α_i we recover (2.91), provided that the necessary independence assumptions are satisfied.

One important characteristic of the model (2.91) is that if the individuals are divided into two or more groups on the basis of ability and the item difficulties estimated separately from each group, then the same $\{\beta_j\}$ are obtained, except possibly for a constant shift and, of course, random errors of estimation. This is called the property of specific objectivity. For some purposes it is too stringent or may be inconsistent with the data. One generalization is to allow the logistic response curves to have different slopes as well as different locations, giving instead of (2.92)

$$e^{\alpha \gamma_j + \beta_j}/(1 + e^{\alpha \gamma_j + \beta_j}) \tag{2.93}$$

and to the corresponding generalization of (2.91).

Various extensions of the model are possible, for example to deal with multiple choice questions in which individuals with large negative α may nevertheless have a positive probability of guessing the correct answer.

The likelihood for the model (2.91) is

$$\exp\left(\sum \alpha_i y_{i.} + \sum \beta_j y_{.j}\right)/\Pi(1 + e^{\alpha_i + \beta_j}), \tag{2.94}$$

showing that under this model only the row and column totals are relevant. Note that this is not true for (2.93), where γ_j, if known, could be regarded as the weight or score to be attached to a correct answer to the jth question. Under the simplest model, however, only the unweighted total number of correct answers is relevant in assessing a particular individual.

For some purposes of comparison of individuals it is thus enough to work with total scores, although for others it may be required to estimate the $\{\alpha_i\}$, or correspondingly to estimate the item easiness parameters $\{\beta_j\}$. Unconditional maximum likelihood would proceed directly from (2.94), although conditional maximum likelihood is preferable. For this, note that for inference about the $\{\beta_j\}$, we write $y_{i\cdot} = \sum_i y_{ij}$, $y_{\cdot j} = \sum_j y_{ij}$ for the row and column totals and divide the full likelihood

$$\exp\left\{\sum_i \alpha_i y_i + \sum_j \beta_j y_j\right\}\bigg/\prod_{i,j}(1 + e^{\alpha_i + \beta_j})$$

by the marginal likelihood of the row totals, $y_{i\cdot} = \sum_j y_{ij}$, to obtain the conditional log likelihood

$$\sum_j \beta_j y_j - \log\left\{\sum_i \sum_{k\in\mathscr{S}_i} e^{\beta_k}\right\}, \tag{2.95}$$

where $\mathscr{S}_i$ is the set of all distinct selections of $y_{i\cdot}$ columns from $\{1,\ldots,c\}$.

Testing the adequacy of the model (2.91) can proceed relatively formally by fitting an extended model such as (2.93), or by dividing the individuals into groups by some external criterion such as age or sex and testing the equality of the item parameters in the two groups. More exploratory methods can be based on splitting the data according to the answer to one of the questions and plotting against one another the estimated difficulties in the two groups (Molenaar, 1983).

2.11 Binary time series

Underlying virtually all the previous discussion lies a quite strong assumption of independence. When, as is quite often desirable, this assumption is abandoned, appreciable complications are to be expected. In the present section we sketch some of the considerations in the analysis of binary time series. That is, we suppose available for analysis a binary response recorded at a series of equally spaced time points; denote the observations by $\{y_1,\ldots,y_n\}$. We deal first with the situation where one or more individual series are given, interest being focused on possible dependencies. Rather different considerations apply for a binary regression problem with possible serial dependency between successive trials, interest being focused on the regression coefficients themselves.

The assumption that the observations are equally spaced in time is important. If the departures from equal spacing are serious enough that they must be allowed for, it may be good to formulate a model in continuous time of which the observations are a sampled or aggregated form. Thus in a medical trial a patient's status might be assessed 1, 3, 6, 12, 24 months after treatment; presence or absence of rain might be recorded in periods of very variable length. From now on, however, we suppose that any departure from equal spacing can be ignored.

Simple descriptive methods will often be desirable as a preliminary to more detailed analysis and may sometimes provide an adequate final analysis. Thus to look for possible trends we may divide the data into a fairly small number of sections and calculate the proportion of successes in each; to look for a cyclic effect of pre-assigned wavelength, we may group the data based on the phase of the proposed cycle, e.g. by month of the year where an annual cycle is suspected.

For examining short-term dependency, study of one-step or higher-order transition counts is natural. A one-step transition count r_{uv} is

Table 2.17 *Artificial example of transition counts: (a) data; (b) one-step counts and proportions, (c) higher-order properties*

(a) 0 0 1 1 1 1 0 0 0 0 0 0 1 1 0 0 1 0 0 0 0 0 0 0 1 1 1

(b) One-step r_{uv}

	Counts		Proportions	
	0	1	0	1
0	12	4 | 16	0.750	0.250
1	3	6 | 9	0.333	0.667
	15	10		

(c) Two-step

	Counts		Proportions	
	0	1	0	1
0	9	7 | 16	0.562	0.438
1	5	3 | 8	0.625	0.375
00	8	4 | 12	0.667	0.333
01	1	3 | 4	0.250	0.750
10	3	0 | 3	1.000	0.000
11	2	3 | 5	0.400	0.600

defined as the number of times $y_i = u$ is followed by $y_{i+1} = v$. There are four such numbers corresponding to $u, v = 0$, 1 and they can be arranged in a 2×2 table; Table 2.17 illustrates the calculation for a simple artificial set of data. From the transition counts we can calculate transition proportions, normally working forward in time, i.e. $p_{uv} = r_{uv}/(r_{u0} + r_{u1})$ is the proportion of times u is followed by v.

Note that the row and column sums of the transition counts are equal except for end effects. In fact

$$r_{00} + r_{01} + (1 - y_n) = r_{00} + r_{10} + (1 - y_1) = r_0,$$
$$r_{10} + r_{11} + y_n = r_{10} + r_{11} + y_1 = r_1,$$

where r_0 and r_1 are the total numbers of 0s and 1s. The end effects can be removed by a circular definition, i.e. by defining $y_{n+1} = y_1$, but this is in most practical contexts artificial and to be avoided.

To study higher-order dependencies transition counts can be defined in two ways. A direct generalization of the one-step count is just to count $r_{uv}^{(2)}$, the number of times $y_i = u$ is followed two steps later by $y_{i+2} = v$. Provided there is sufficient data a more enlightening approach is often to consider $r_{uv,w}$, the number of times the pair u, v is followed by w; again see Table 2.17. This leads to the study of four proportions, namely the proportion of 1s following respectively 00, 01, 10, 11. Of course this more detailed investigation fairly speedily becomes impracticable if we look at higher-order dependencies.

As with other time series problems it is essential for success that the analysis is adapted to the time scale of the phenomenon under study. Thus it is ineffective to look for long-term trends or slow cycles via one-step transition counts. Nor are methods based on transition counts suitable for analysing long series of 0s interspersed with occasional clusters of 1s.

For more formal analysis we turn first to a simple homogeneous two-state Markov chain, this being the simplest process showing dependency between successive trials. Such a model is defined by the Markov property and by a matrix of transition probabilities

$$\pi = \begin{bmatrix} \pi_{00} & \pi_{01} \\ \pi_{10} & \pi_{10} \end{bmatrix} \tag{2.96}$$

where $\pi_{00} + \pi_{01} = \pi_{10} + \pi_{11} = 1$ and

$$\pi_{uv} = \text{prob}(Y_{i+1} = v \mid Y_i = u).$$

To calculate a log likelihood function we use the Markov property to give in a self-explanatory notation

$$l(y_1) + l(y_2|y_1) + l(y_3|y_2) + \cdots + l(y_n|y_{n-1}). \qquad (2.97)$$

If it were not for the Markov property each term would have to be conditioned on all previous values, not just on the immediately preceding value.

All terms except the first are given directly from the matrix π. For the first term there are two possibilities. One is to treat y_1 as a given constant, therefore contributing merely a unit factor to the likelihood. The second is, where reasonable, to give Y_1 the equilibrium distribution of the chain, i.e. to suppose that data collection starts at an arbitrary time a long way after the start of the process. For a single long series the additional information contributed in the second approach is negligible, but where a large number of individually short series are under analysis it will be appreciably more efficient to use the second approach, although it will be a wise precaution to check that the observed proportion of series for which $y_1 = 1$ is consistent with that to be expected from the other part of the series.

The log likelihood for fixed y_1, which we denote by l_c, is easily seen to be

$$l_c = \sum r_{uv} \log \pi_{uv}, \qquad (2.98)$$

so that the matrix of transition counts is sufficient for the parameters π. The log likelihood is exactly that for sampling a multinomial distribution and the maximum likelihood estimates of the individual π_{uv} are the corresponding proportions. We can use the logit difference $\Delta = \log(\pi_{11}/\pi_{10}) - \log(\pi_{01}/\pi_{00})$ as a reasonable measure of dependence in the Markov chain; to a very close approximation all the methods for 2×2 tables can be applied here, including methods for amalgamating information from different individuals presumed to have a constant Δ, even if different matrices π.

If, however, we use the second form of condition for y_1, we add to l_c the term

$$y_1 \log \pi_{01} + (1 - y_1) \log \pi_{10} - \log(\pi_{01} + \pi_{10}). \qquad (2.99)$$

If we have a large number of independent series each of length n a simple calculation yields the Fisher information matrices for (π_{01}, π_{10}) based on the two forms of likelihood. Especially in chains with appreciable persistence there may be important gains in precision from using the initial conditions.

Similar arguments apply when m-dependent chains are used instead of a simple (one-dependent) Markov chain. As in Table 2.17, we have in effect a binary response depending on an explanatory variable which for $m = 2$ takes the four values $(0, 0), \ldots, (1, 1)$. One use is to test consistency with a simple Markov chain by comparing maximized log likelihoods under the two fits; note, however, that the same set of response variables must be used in the two cases, so that in fitting the simple model the contribution from both y_1 and y_2 should be omitted.

As m increases the number 2^m of preliminary conditioning sequences increases rapidly and it is natural to look for simpler representations, one such being the 'main effect' model in which

$$\lambda_i = \log \{\text{prob}(Y_i = 1)/\text{prob}(Y_i = 0)\} = \alpha + \beta_1(Y_{i-1} - \tfrac{1}{2}) + \cdots$$
$$+ \beta_m(Y_{i-m} - \tfrac{1}{2}). \tag{2.100}$$

Example 2.19 Possible clustering of point events
We describe in outline an application in which models of the above type can be used to assess possible clustering or other structure in a series of binary events. Data were available from a custodial centre for young offenders showing for each day whether an abscondment had occurred that day or not; further data showed the number of boys absconding but we disregard that for the present discussion. For day i let $Y_i = 1$ if an abscondment takes place, with $Y_i = 0$ otherwise. Now it is likely, and was confirmed by inspection of the data, that there are systematic effects depending on the day of the week and on the time of the year. Therefore models were fitted of the general form

$$\lambda_i = \alpha + \beta_1 Y_{i-1} + \text{day of week term}$$
$$+ \text{month of year term},$$
$$\lambda_i = \alpha + \beta_1 Y_{i-1} + \beta_2 Y_{i-2} + \text{day of week term}$$
$$+ \text{month of year term},$$

and so on. The first set of results examined gave a small and negative β_1 and a larger and positive β_2, suggesting that an abscondment was relatively likely to be followed by another two days later. This effect having been found in one institution, it was important to see whether it was confirmed at other similar institutions, but when this was examined no systematically consistent effects were found.

The above models are observation driven in the sense that the probability at time i is determined by the *observations* at previous times. Thus in Example 2.19 it is supposed that the probability of an abscondment on a particular day is influenced by whether there was an abscondment the previous day. Sometimes, however, it is more reasonable to consider an unobserved process of probabilities with some simple structure such that this process is sampled independently at each time point. Such processes are called parameter driven; they are closely related to the state-space models of time series theory. Unfortunately such models tend to be difficult to handle as a basis for the analysis of data. Parameter driven as contrasted with observation driven models are quite widely used in learning theory.

The simplest model of this type is obtained by supposing the logit transform λ_i of the probability of success on the ith trial evolves in accordance with a Gaussian first-order autoregression

$$\lambda_{i+1} = \mu + \rho(\lambda_i - \mu) + \varepsilon_{i+1},$$

where the innovation process $\{\varepsilon_i\}$ is a Gaussian sequence of independent and identically distributed random variables of zero mean and variance σ_ε^2. We suppose that conditionally on the unobserved sequence $\{\lambda_i\}$ the observations $\{Y_i\}$ are mutually independent binary random variables Y_i having probability of success determined only by λ_i.

Such models in general require computer-intensive methods for their study, although some insight into their behaviour can be obtained via the approximation in which the linear logistic model is replaced by a linear model in the probabilities themselves. The second-order structure of the process is then, as is to be expected, that of a first-order autoregression with error of observation, and so decays geometrically except for the deflated first term; this can in principle be checked via the higher-order transition counts.

A different type of model results if it is reasonable to postulate an unobserved stationary Gaussian process $\{\Gamma_i\}$, with $Y_i = 1$ if and only if $\Gamma_i > 0$. This is, of course, closely related to the idea of tolerance distribution discussed for individual responses in Section 1.3. We can without loss of generality take $\text{var}(\Gamma_i) = 1$; denote its mean by μ and autocorrelation function by $\rho_\Gamma(h) = \text{corr}(\Gamma_i, \Gamma_{i+h})$. Now μ can be estimated by equating $\sum y_i/n$, the observed proportion of 1 s to $\Phi(\mu)$. Further $\rho_\Gamma(h)$ can be estimated from the h-step transition count via the estimating equation for

$$r_{11}^{(h)}/n = \Phi_2(\mu, \mu; \tilde{\rho}),$$

where the right-hand side is the probability in a bivariate normal distribution of mean (μ, μ), unit variances and correlation $\tilde{\rho}$ lying in the positive quadrant. While no formal study of efficiency is available in general, it is now possible in principle to estimate the underlying correlation function and thence, if desired, to fit one of the 'standard' time series models. One special case which is simpler and which has been studied in detail by Kedem (1980) assumes that $\mu = 0$, i.e. that the marginal proportion of 1s is $\frac{1}{2}$. This is likely to be reasonable only when the originating continuous series is available and the cut-off defining $y = 1$ determined as, for instance, the overall median. Such an analysis would be useful if, for example, the continuous series were assumed to be some unknown transformation of a Gaussian series and if it were required to express the structure of the process on the Gaussian scale.

Finally we turn to situations in which interest lies not so much in the time series structure itself as in a regression problem in which autocorrelation is suspected in the random variability, the regression coefficients being the objective of study. The simplest problem of this kind arises when data are available on a number of independent individuals in the form of a time series of responses, the explanatory variables being constant for each individual.

In such situations it will often be sensible not to attempt detailed representation of the time series structure within an individual, but rather to condense each individual's responses into a small number of meaningful summary statistics, for example overall proportion of successes, a summary measure of one-step dependence, of trend, cyclic behaviour and so on. These can then be analysed as derived responses, the error of estimated regression coefficients being determined from the between-subject variability.

If, however, the explanatory variables change within subjects, the situation is much more complicated. When the individual sequences are relatively long it may again be feasible to proceed informally, but if they are short analysis of the probability model is likely to be especially fruitful. We shall not explore this further.

Bibliographic notes

Formal analysis of the 2×2 table has an extraordinary history. Fisher's exact test, and the associated interval estimation theory, is best approached via Fisher (1935) and the test is very close to chi-squared with continuity correction (Yates, 1934). Tocher (1950) in effect showed that this gave optimal p-values, although that was not his terminology. If, however, preassigned levels, e.g. 0.05, were to be required, then randomization, or something essentially equivalent, is called for. There has been a long and continuing controversy on this; see Yates (1984) and the ensuing discussion. Pearson (1947) and Barnard (1947) clarified the various distinct schemes leading to the 2×2 table.

The saddle-point approximation to the conditional log likelihood function for the 2×2 table is derived by Barndorff-Nielsen and Cox (1979); a more general result is presented by Davison (1988). If only the conditional maximum likelihood estimate is required, an alternative approximation based on a result of Mantel and Hankey (1975) is given in McCullagh and Nelder (1983, p. 90); it has been shown to give estimates close to the correct value (Breslow and Cologne, 1986).

For the combination of several or many 2×2 tables see Cochran (1950), Cox (1958a) and especially Mantel and Haenszel (1959).

The conditional and unconditional estimates of the common odds ratio in several 2×2 tables are asymptotically equivalent if the number of tables remains fixed (Gart, 1970, 1971; Breslow, 1976). However, the unconditional estimate can be badly biased if the number of tables is large, especially if the data are sparse (Pike, Hill and Smith, 1980; Breslow and Day, 1980, Chapter 7; Lubin, 1981).

Tarone (1985) considers testing the heterogeneity of the odds ratio and derives an expression which is based on score statistics and involves a correction to the usual goodness-of-fit statistic.

The widely used Mantel–Haenszel estimator is shown by Breslow (1981) to be consistent even with sparse data. Estimation of its variance is considered by Hauck (1979) and Robins, Breslow and Greenland (1986). Simulation studies show that the Mantel–Haenszel estimator has high efficiency under most conditions (Anderson, Auquier, Hauck et al., 1980, Chapter 7; Hauck, 1984; Jewell, 1984; Donner and Hauck, 1986; Hauck and Donner, 1988). Donner and Hauck (1988) propose a modified Mantel–Haenszel estimator for use when dependence between responses invalidates the binomial assumption; see also Liang (1985).

Estimation of the common relative risk, rather than the odds ratio, is considered by Gart (1985).

Experimental design for the comparison of binomial proportions is reviewed by Abdelbasit and Plackett (1981); see also Minkin (1987).

The significance test for matched pair data is due to McNemar (1947), and the deduction from a logistic model and the associated estimation procedure to Cox (1958b). Altham (1969) gave a Bayesian formulation. For closely related problems, Schaafsma (1973), in effect, advocated concentration on the probabilities of the various patterns of response that could arise, as contrasted with not directly testable modelling. The extension to a logistic model for a cross-over trial was given by Gart (1969). Recently there has been considerable interest in the possibility of improved inference by in effect recovering information from between pair differences; for a thorough account and review, see Kenward and Jones (1987, 1989) and for a different development Fidler (1984). For the special models needed to analyse mixtures of drugs, see Hewlett and Plackett (1964), and Plackett and Hewlett (1967).

Deviance residuals and diagnostics for binary data similar to those for normal-theory models were introduced by Pregibon (1981). Deletion diagnostics are also considered by Cook (1986). Pierce and Schafer (1986) discuss deviance and Pearson residuals. Williams (1987) recommends a residual which combines both definitions. Cox and Snell (1968, 1971) define a residual aimed to be more normally distributed. Methods based on score statistics are considered by Pregibon (1982) and Jørgensen (1983). Partial residuals are derived and plotted by Landwehr, Pregibon and Shoemaker (1984). Partial residual plots after smoothing are recommended by Fowlkes (1987). Wang (1985) considers added variable plots. Hastie and Tibshirani (1986, 1987a, b) adopt a non-parametric approach to detect the form for an explanatory variable.

For some general discussion on residuals and diagnostics, see McCullagh and Nelder (1983, Section 2.4 and Chapter 11).

The application of logistic models to large contingency tables is discussed by Fowlkes, Freeny and Landwehr (1988).

Methods of fitting resistant to outliers are considered by Green (1984) and Copas (1988).

There is a close connection between the analysis of factorial experiments with a binary response and the study of multidimensional contingency tables, although in the latter the various dimensions are usually treated symmetrically. The distinction between linear and log linear analyses is reviewed by Darroch (1962, 1974) and Imrey, Koch and Stokes (1981, 1982); see also Palmgren (1981). Most recent work has concentrated on log linear models (Bishop, Fienberg and Holland, 1975; Fienberg, 1977; Plackett, 1981; Goodman, 1985).

For Simpson's paradox, see Simpson (1951) and Blyth (1972).

A good introduction to the Rasch (1960) model is in the book of Andersen (1980). Goodness of fit is discussed by Andersen (1973b) and Molenaar (1983).

For binary time series treated as Markov chains, see Billingsley (1961) for a lucid review of key results. The extensive recent literature on inference in stochastic processes is, of course, relevant; see Basawa and Scott (1983). The treatment via hard limited Gaussian processes is discussed in detail by Kedem (1980). For some recent work involving regression models with correlated errors, see Azzalini (1983), West, Harrison and Migon (1985) and Liang (1985).

CHAPTER 3

Some complications

3.1 Introduction

In the previous chapter we have ranged over a number of types of application. All the work except the section on time series has concerned independent observations on binary data in which the probability of success changes in some relatively simple fashion, represented in terms of smooth dependence on unknown parameters, the estimation of which is regarded as the primary focus of the analysis.

This relatively simple formulation can need elaboration in various ways. We now consider three such. First and perhaps most importantly there is the possibility that representations in terms of independent binary random variables with simply varying probabilities underestimate the random fluctuations that are present. In Section 3.2 we consider a number of ways in which such overdispersion could arise. Then we consider briefly the related matter of empirical Bayes estimation, i.e. of 'borrowing strength' about the value of one parameter from data on related parameters. Finally, in Section 3.4 we examine the effect of measurement errors in either response or explanatory variables.

3.2 Anomalous dispersion

3.2.1 General remarks

As noted above, except in the discussion of time series models, independence of individuals has been assumed throughout the previous chapters. In particular if $E(Y_j) = \theta$ $(j = 1, \ldots, m)$, then $E(\sum Y_j) = m\theta$ and for binary random variables

$$\text{var}(Y_j) = E(Y_j^2) - \{E(Y_j)\}^2 = \theta(1 - \theta).$$

Thus, provided only that the different Y_j's are pairwise independent, we have without further assumption that if $R = \sum Y_j$ is the total number of successes, then

$$\text{var}(R) = \sum \text{var}(Y_j) = m\theta(1 - \theta). \tag{3.1}$$

That is, there is a mathematical relation between the variance and the mean of the number of successes in m trials holding as a consequence solely of the assumed constancy of the probability of success and of the independence of the trials.

Now this relation can be tested empirically, either directly by having groups of m individuals all of whom it is reasonable to regard as having constant θ, or indirectly by examining dispersion from a broadly well-fitting model. When such comparisons are made it is relatively common to find overdispersion, i.e. to find that the empirical variance for a given θ exceeds (3.1). Underdispersion is sometimes observed, although this is uncommon.

Overdispersion is usually handled by regarding it as an additional source of random variability, or of positive correlation between the responses within a group, although in principle it suggests the existence of further explanatory variables whose inclusion in the model would reduce the variance to the 'standard' form (3.1). Underdispersion is accounted for by negative correlation between the individuals in a group, induced for instance by competition, or by systematic variation in the probability of success.

There are several consequences of, for example, overdispersion. Sometimes one may be specifically interested in the overdispersion as a phenomenon in its own right, in which case its magnitude needs to be estimated and sometimes even quite detailed analysis of its structure may be called for. Related to this is the use of empirical Bayes procedures for improving the estimate of the probability in one 'group' by 'borrowing strength' from related groups.

More commonly, however, interest remains in the systematic effect of explanatory variables. Then there are two considerations. First, there may be a loss of efficiency in the point estimates if overdispersion is ignored. That is, estimates of regression coefficients different from the standard ones may be preferable. On the whole this is not in practice a serious matter. Secondly, and much more importantly, even if we keep to the 'usual' point estimates, their precision will be degraded by overdispersion, so that standard errors calculated ignoring overdispersion will be too small. Failure to correct for this can be seriously misleading.

If the amount of overdispersion is large, i.e. if the component (3.1) is a relatively small part of the total variation of R, it may be best to concentrate directly on the form of the variation of the total number of successes, disregarding its structure as a sum of binary contributions.

3.2.2 Internal dependence

As explained above, dependence between the responses of different individuals in a group is one explanation or representation of overdispersion. Detailed study of such dependence may sometimes be worthwhile, but here we concentrate on some simple possibilities.

First, suppose that all pairs of individuals have the same joint distribution for their binary responses (Y_i, Y_j). This is the analogue of the so-called intra-class correlation model in normal theory. Write for $i \neq j$

$$\left\{ \begin{array}{l} E(Y_i) = E(Y_j) = \theta, \quad \text{prob}(Y_i = Y_j = 1) = \theta^2 + \theta(1 - \theta)\rho, \\ \text{prob}(Y_i = Y_j = 0) = (1 - \theta)^2 + \theta(1 - \theta)\rho, \\ \text{prob}(Y_i = 0, \ Y_j = 1) = \text{prob}(Y_i = 1, \ Y_j = 0) = \theta(1 - \theta)(1 - \rho). \end{array} \right\} \quad (3.2)$$

An immediate restriction on ρ is that

$$-\min\{\theta/(1-\theta), \quad (1-\theta)/\theta\} \leqslant \rho \leqslant 1. \quad (3.3)$$

It follows from (3.2) that $\text{cov}(Y_i, Y_j) = \theta(1 - \theta)\rho$, so that

$$\begin{aligned} \text{var}(R) &= m\theta(1 - \theta) + m(m - 1)\theta(1 - \theta)\rho \\ &= m\theta(1 - \theta)[1 + (m - 1)\rho]. \end{aligned} \quad (3.4)$$

Note that a further restriction on ρ arises if (3.2) is to apply to all pairs chosen from $Y_1, \ldots, Y_m$, namely that $\rho \geqslant -(m - 1)^{-1}$.

If ρ is constant as θ varies, then (3.4) represents over- or underdispersion by a fixed factor; it may sometimes help in interpreting the observed dispersion to obtain an estimate of ρ. Note also that if we have to analyse data in which group size m varies, (3.4) predicts how the observed dispersion should change with m.

Another simple form of dependence arises when only adjacent pairs are correlated, i.e. (3.2) applies with $j = i - 1, \ i + 1$, but otherwise different binary responses are pairwise independent. Then

$$\text{var}(R) = m\theta(1 - \theta)[1 + 2(1 - 1/m)\rho] \quad (3.5)$$

in contrast with (3.4). Yet another possibility is to represent $\{Y_1, \ldots, Y_m\}$ as a stationary Markov chain.

'Competition' between individuals within a group will induce negative correlation which may be crudely represented by (3.4) or (3.5); we shall not consider more detailed representations.

3.2.3 Variable probabilities

A rather different study of dispersion changes is obtained by retaining independence of the component binary random variables but abandoning the assumption in (3.1) that θ is constant. This can be done in two different ways, leading respectively to under- and overdispersion.

First suppose that $E(Y_j) = \theta_j$ $(j = 1, \ldots, m)$. That is, the probability of success varies within each set of m trials in a way that is exactly reproduced on repetition. Then

$$\operatorname{var}(R) = \sum \operatorname{var}(Y_j)$$
$$= \sum \theta_j(1 - \theta_j) \;=\; m\bar\theta - \Sigma\theta_1^2$$
$$= m\bar\theta(1 - \bar\theta) - \sum(\theta_j - \bar\theta)^2, \qquad (3.6)$$

where $\bar\theta = \sum \theta_j/m$. Thus if $\sum(\theta_j - \bar\theta)^2 = m\bar\theta(1 - \bar\theta)v$,

$$v = \frac{\Sigma(\theta_j - \bar\theta)^2}{m\bar\theta(1-\bar\theta)}$$

$$\operatorname{var}(R) = m\bar\theta(1 - \bar\theta)(1 - v) \qquad (3.7)$$

and observed underdispersion can be converted into an estimate of v. Again, if sets of data are available for various θ the underdispersion is by a constant factor if and only if v is constant as θ varies.

As an extreme example suppose that m is even and that half the θ_j are zero and the other half one, so that $\bar\theta = \frac{1}{2}$. The variance of the total number of successes is zero and in this extreme case it is easily seen that $v = 1$.

A quite different form of non-constancy of the probability of success arises when θ is constant for all the individuals in a group but varies between groups, i.e. between replications. To study this possibility suppose that θ is the realized value of a random variable Θ with a probability density function $f(\theta)$ $(0 \leqslant \theta \leqslant 1)$. Then

$$E(R \mid \Theta = \theta) = m\theta, \qquad \operatorname{var}(R \mid \Theta = \theta) = m\theta(1 - \theta),$$

and on considering the unconditional mean and variance, we have that

$$E(R) = m\mu_\theta,$$

(marginal handwritten note:) $= m\bar\theta - m\bar\theta - m\bar\theta^2 \quad \Sigma\theta_1^2 + 2m\bar\theta^2 \quad m^2\bar\theta^2$

$$\operatorname{var}(R) = E\{m\Theta(1 - \Theta)\} + \operatorname{var}(m\Theta)$$
$$= m\mu_\theta(1 - \mu_\theta) + m(m - 1)\sigma_\theta^2, \tag{3.8}$$

where μ_θ and σ_θ^2 are the mean and variance of the distribution of Θ. The condition that there is an inflation of variance by a factor independent of μ_θ is thus that

$$\sigma_\theta^2 = \gamma\mu_\theta(1 - \mu_\theta),$$

when

$$\operatorname{var}(R) = m\mu_\theta(1 - \mu_\theta)\{1 + (m - 1)\gamma\}, \tag{3.9}$$

this being formally identical to (3.4), although (3.4) is interpretable for negative as well as for positive ρ. If it were required to study in more detail the dependence of σ_θ^2 on μ_θ it would be reasonable to write

$$\sigma_\theta^2 = \gamma_0\mu_\theta^{\gamma_1}(1 - \mu_\theta)^{\gamma_2} \tag{3.10}$$

and to attempt estimation of γ_1 and γ_2.

In principle the exact distribution of $R = \sum Y_j$ can be determined under this random compounding model. In particular if Θ has the beta density

$$\theta^{\alpha - 1}(1 - \theta)^{\beta - 1}/B(\alpha, \beta),$$

we have that

$$\operatorname{prob}(R = r) = \int_0^1 \operatorname{prob}(R = r | \Theta = \theta) f(\theta) d\theta$$

$$= \binom{m}{r} \int_0^1 \theta^{r + \alpha - 1}(1 - \theta)^{m - r + \beta - 1} d\theta/B(\alpha, \beta)$$

$$= \frac{\Gamma(m + 1)\Gamma(r + \alpha)\Gamma(m - r + \beta)\Gamma(\alpha + \beta)}{\Gamma(r + 1)\Gamma(m - r + 1)\Gamma(m + \alpha + \beta)\Gamma(\alpha)\Gamma(\beta)}, \tag{3.11}$$

which is called the beta-binomial distribution.

The mean and variance of the beta distribution are such that

$$\mu_\theta = \alpha/(\alpha + \beta), \qquad \gamma = \operatorname{var}(\Theta)/\{\mu_\theta(1 - \mu_\theta)\} = (\alpha + \beta + 1)^{-1}.$$

For some purposes the most natural parameterization of the distribution is in terms of μ_θ, or equivalently α/β, and γ, or equivalently $\alpha + \beta$. It follows from (3.8) that

$$E(R) = m\alpha/(\alpha + \beta) = m\mu_\theta,$$
$$\operatorname{var}(R) = m\mu_\theta(1 - \mu_\theta)\{1 + (m - 1)\gamma\}. \tag{3.12}$$

3.2.4 Quasi-likelihood estimation

When interest is focused on regression parameters in a linear logistic model for the marginal probabilities and the under- or overdispersion is by a constant factor as in (3.4), (3.5), (3.7) and (3.9), it is reasonable to fit by the ordinary maximum likelihood procedure for the standard model: use of maximum likelihood procedures in this way when the systematic structure is 'correct' but the error structure under- or overdispersed by a suitable factor is called quasi-likelihood estimation. In the simplest case when all the groups have the same size, m, we have

$$\operatorname{var}(R) = m\tau\theta(1 - \theta); \tag{3.13}$$

then variances of estimates and corresponding test statistics have to be increased by a factor τ over their ordinary maximum likelihood values, and to do this τ has either to be estimated or a value assigned to it on the basis of previous experience. The former can be done via a suitable goodness of fit statistic T, usually a Pearson chi-squared statistic having a chi-squared distribution with, say, q degrees of freedom under the standard assumptions and with expectation τq under the modified model; thus T/q estimates τ. If the group sizes are different, the simplest natural modification to (3.13) is to assume that

$$\operatorname{var}(R_j) = m_j\{1 + (m_j - 1)\gamma\}\theta_j(1 - \theta_j).$$

The parameter γ has now to be estimated iteratively. Note, however, that most weighted estimation procedures are relatively insensitive to the weights, so that unless the variation in the overdispersion factor is large or the parameter γ of intrinsic interest a simpler analysis based on a single overdispersion factor is likely often to be adequate.

Such an analysis is simple, approximate and usually somewhat inefficient; however, the approximations and inefficiency can be removed by a more detailed specification of the nature of the dispersion, a specification that would often be difficult to achieve realistically in practice. The main practical limitation on the quasi-likelihood analysis is the assumption that the change in variance is by a simple factor throughout.

The key ideas are best illustrated by the simplest special case.

3.2.5 A simple example

Consider $R_1, \ldots, R_k$, these being the numbers of successes in m trials, it being assumed that there is no systematic variation in the probability of success. The quasi-likelihood procedure in this special case is to estimate θ by $\hat{\theta} = \sum R_j/(mk) = \bar{R}/m$, the observed proportion of successes, and the variance of this by

$$\tilde{\tau}\tilde{\theta}(1 - \tilde{\theta})/(mk), \tag{3.14}$$

where $\tilde{\tau}$ is an estimate of the over dispersion factor.

Now to estimate τ we need a goodness of fit statistic for the homogeneous binomial model. An obvious one is

$$T_p = \sum (R_j - \bar{R})^2/\{\bar{R}(m - \bar{R})/m\}. \tag{3.15}$$

Under the more general model this has expectation approximately $\tau(k - 1)$, leading to an estimate $\tilde{\tau}$ and to an estimate of the variance of $\tilde{\theta}$ via (3.14).

The resulting estimates of θ and standard error are thus exactly those resulting from treating $R_1, \ldots, R_k$ as a random sample from an arbitrary distribution. That is, the variance of R is estimated as

$$\sum (R_j - \bar{R})^2/\{k(k - 1)\}.$$

If we focus interest on the component of variance σ_θ^2 we can construct via the above expressions the exactly unbiased estimate

$$\sum (R_j - \bar{R})^2(km - 1)/[m^2(m - 1)k(k - 1)]$$
$$- \bar{R}(m - \bar{R})/[m^2(m - 1)], \tag{3.16}$$

although for a single set of data the estimate is taken in the usual way to be zero if (3.16) is negative.

For a technically more efficient analysis a more detailed specification is needed, the simplest being the beta-binomial model of Section 3.2.4 to which a full maximum likelihood analysis can be applied. The asymptotic efficiency of $\bar{R}/m$ relative to the maximum likelihood estimate as an estimate of μ_θ or equivalently α/β exceeds 0.93 unless α and β are both less than one. The efficiency of the estimate (3.16) as an estimate of σ_θ^2, or equivalently $\alpha + \beta$, is lower (Ruiz, 1989).

Unless specific interest attaches to the form of the overdispersion the simpler, more widely based, method seems preferable. We shall return to this point in connection with empirical Bayes estimation.

3.2.6 A more complex example of overdispersion

The situation discussed in Section 3.3.5 can be generalized in various ways. First, if the sample sizes in the different groups are different, i.e.

R_j is the number of successes in m_j trials, then complications essentially those arising in unbalanced one-way component of variance analysis arise. Thus μ_θ can be estimated via $\sum R_j / \sum m_j$ or via $k^{-1} \sum R_j / m_j$, as well, of course, via intermediate weighted estimates. We shall not discuss details.

A more major generalization arises when there is much more complicated structure, possibly partly systematic and partly random, present in the data. There are a large number of possible approaches, depending in part on the linearizing transformation, logit, probit,... used to achieve a simple representation.

In simple balanced arrangements in which also the probability of success is never extreme, a direct linear model may be the best approach. Thus in a row × column arrangement with an equal number m of observations per cell it may be reasonable to write in a self-explanatory notation

$$\theta_{ij} = \mu + A_i + B_j, \tag{3.17}$$

where if both 'row' and 'column' effects are random the $\{A_i\}$ and $\{B_j\}$ are uncorrelated sets of random variables of zero mean, with variances respectively σ_A^2 and σ_B^2. Estimates can be formed in almost the usual way from the sums of squares between rows and between columns.

In many cases, however, a direct linear model is unlikely to be satisfactory, for the reasons that have already been discussed. The most effective approach of any generality applies when the data are grouped into sets within each of which simple conditions of independence and constant probability are likely to hold and such that the relevant probabilities are described by an additive model on a suitable scale, e.g. the logistic, that model involving systematic and random terms. Then, approximately, application of an empirical logistic or other transform leads to an unbalanced standard component of variance problem on which there is a large literature.

One particular motivation of such a model comes from the latent variable approach sketched in Section 2.1. If the latent variable follows the natural normal-theory model for the type of structure in question, we may take the lowest variance component as unity and the threshold for $Y = 1$ as zero and deduce therefrom a probit model. Analysis will be difficult, however, unless grouping is possible, when an empirical inverse normal transformation will produce new observations to which standard methods can be applied, approximately.

3.2.7 Two examples

We now give two examples illustrating the need to consider overdispersion in analysing binary data.

Example 3.1 Fertilizer experiment on growth of cauliflowers
(*continued*)
In Example 2.16 we fitted a logistic model to data from a factorial experiment in which the response in each cell is the proportion of cauliflowers out of $m = 48$ that were grade 24 or better; the variability present was greater than that specified by a simple binomial model. Applying the quasi-likelihood procedure of Section 3.2.5, we compute the goodness of fit statistics generalizing (3.15), $T_P = 20.54$, and multiply the standard errors of the estimated coefficients by $\sqrt{(T_P/8)} = 1.60$ (see Table 3.1). Both the linear and quadratic components of nitrogen, N_L and N_Q, are significant. The effect of potassium is not significant.

Example 3.2 Germination of seeds in a 2 × 2 factorial experiment
(Crowder, 1978)
Table 3.2 gives the number of seeds germinating in a 2 × 2 factorial experiment in which two types of seed and two root extracts, bean and cucumber, were compared. Seeds were brushed on to a plate covered with root extract and the number subsequently germinating or not germinating were counted.

Crowder (1978) fitted a beta-binomial model (3.11). We consider a quasi-likelihood approach (Williams, 1982) and fit a logistic model with parameters representing the factors of the 2 × 2 experiment.

Table 3.1 *Growth of cauliflowers. Fitted logistic model*

Effect	Estimate	Adjusted standard error
Constant	0.092	0.125
K	−0.042	0.125
N_L	0.296	0.058
$K \times N_L$	0.024	0.058
N_Q	0.266	0.125
Blocks: $K \times N_Q$	−0.132	0.125
R	−0.411	0.125
$R \times K \times N_Q$	−0.033	0.125

Table 3.2 *Data for seeds O. aegyptiaco 75 and 73, bean and cucumber root extracts. Number r, out of m, which germinate*

O. aegyptiaca 75				O. aegyptiaca 73			
Bean		Cucumber		Bean		Cucumber	
r	m	r	m	r	m	r	m
10	39	5	6	8	16	3	12
23	62	53	74	10	30	22	41
23	81	55	72	8	28	15	30
26	51	32	51	23	45	32	51
17	39	46	79	0	4	3	7
		10	13				

Source: Crowder, 1978.

Table 3.3 *Estimated probabilities of germination for data of Table 3.2*

	O. aegyptiaca 75		O. aegyptiaca 73	
	Bean	Cucumber	Bean	Cucumber
Model				
Beta-binomial	0.368	0.685	0.391	0.519
Quasi-likelihood	0.369	0.689	0.386	0.511
Logistic	0.364	0.681	0.398	0.532

Since the sample sizes vary from $m = 4$ to $m = 81$ it is necessary to iterate to estimate the parameter γ. The interaction between the two factors, type of seed and root extract, is significant. Estimates of the probabilities of germination under the saturated model are given in Table 3.3; they agree well with those obtained by Crowder.

For comparison we show also in Table 3.3 the probabilities estimated simply from a logistic regression ignoring overdispersion. The estimates are similar to those above, confirming that the method of estimation (iteratively weighted least squares) is relatively insensitive to the variation in the weights.

3.3 Empirical Bayes methods

3.3.1 Preliminaries

We now deal briefly with an interesting collection of ideas which come under the broad heading of empirical Bayes formulations. In these a parameter of interest is regarded as the realization of a random variable having a prior frequency distribution which can be estimated from relevant data. In other words in addition to the data whose distribution depends directly on the parameter of interest, there are other, usually similar, data depending on parameters likely to be fairly close to the parameter of immediate interest.

There are several formulations of this idea; we shall consider only a fully parametric version. Suppose that data are available depending on a general parameter ψ and on parameters $\gamma_1, \ldots, \gamma_k$, where k is fairly large and the γ's are parameters having a similar interpretation and represented by independent and identically distributed random variables with density $p(\gamma; \kappa)$, where κ is an unknown parameter. Suppose that interest focuses on a particular γ, say γ_1, representing some property of the first 'group' of observations; alternatively the parameter of interest may be a contrast, such as $\gamma_2 - \gamma_1$.

If we treat $\gamma_1, \ldots, \gamma_k$ as separate parameters, we obtain an estimate of γ_1. The random generation of $\gamma_2, \ldots, \gamma_k$ means that we can 'borrow strength' from the distribution of $\gamma_2, \ldots, \gamma_k$ to improve the estimate of γ_1. This commonly involves 'shrinking' the direct estimate of γ_1 towards the overall average of all other γ's.

3.3.2 Normal means

It is helpful to begin by discussing estimation for normal means, partly because this shows a very simple form of empirical Bayes estimation and partly because more complicated problems can often be reduced approximately to this one.

Suppose that we have samples of size r from k normal populations of means $\mu_1, \ldots, \mu_k$ and variance σ^2. The corresponding sample means $\bar{Y}_1, \ldots, \bar{Y}_k$ are normal with variance σ^2/r. Suppose now that $\mu_1, \ldots, \mu_k$ are themselves normally distributed with mean ν and variance τ^2 and that interest focuses on μ_1. By Bayes's theorem the posterior distribution of μ_1 is proportional to

$$\frac{1}{\sqrt{(2\pi)}(\sigma/\sqrt{r})} \exp\left\{-\frac{(\bar{y}_1 - \mu_1)^2}{2\sigma^2/r}\right\} \frac{1}{\sqrt{(2\pi)}\tau} \exp\left\{-\frac{(\mu_1 - \nu)^2}{2\tau^2}\right\}.$$

On completing the square of the exponential as a function of μ_1, we have that the posterior density of μ_1 is normal with mean and variance

$$\frac{\bar{y}_1/(\sigma^2/r) + v/\tau^2}{1/(\sigma^2/r) + 1/\tau^2}, \qquad \frac{1}{1/(\sigma^2/r) + 1/\tau^2}. \qquad (3.18)$$

Note that this is the optimally weighted mean of $\bar{y}_1$, the mean of the directly relevant data, and v, the prior mean. If the mean is written as

$$\bar{y}_1 + \frac{1/\tau^2}{1/(\sigma^2/r) + 1/\tau^2}(v - \bar{y}_1), \qquad (3.19)$$

the modification to the sample mean $\bar{y}_1$ is seen as a proportional shrinking towards v.

To use these formulae, v and τ^2/σ^2 have to be estimated, the former from the overall mean $\bar{y}_. = \sum y_j/k$ and the latter via the ratio of mean squares in an analysis of variance. For this, note that we have the standard normal-theory components-of-variance model of analysis of variance under which the expected mean squares between and within groups are respectively

$$\sigma^2 + r\tau^2, \qquad \sigma^2.$$

In fact the distinction between the empirical Bayes and components-of-variance formulations is solely one of emphasis; in the former we are interested, in, say, μ_1 and in the latter in (v, τ^2, σ^2). Provided that k is not too small, e.g. $k > 10$, the effect on (3.18) of estimating v and τ^2/σ^2 is relatively small.

The proportional character of the shrinkage in (3.19) does require approximate normality; long-tailed behaviour implies that extreme observations should not be so drastically shrunk or even not shrunk at all.

3.3.3 Binary data

Suppose now that we have k groups of binary observations, each of m trials, the number of successes being $r_1, \ldots, r_k$.

Suppose for simplicity that interest is focused on the probability in a single group; we can choose the first without loss of generality. We argue as before in two ways. The first uses first and second moment properties, i.e. essentially normal-theory considerations, whereas the second depends explicitly on the beta-binomial formulation.

The former approach can be expressed as follows. We have two independent estimates of θ_1, namely R_1/m and μ_θ. If we combine these, weighting inversely as the variance, we obtain

$$\frac{[(R_1/m)\{\mu_\theta(1-\mu_\theta)/m\}^{-1} + \mu_\theta\sigma_\theta^{-2}]}{[\{\mu_\theta(1-\mu_\theta)/m\}^{-1} + \sigma_\theta^{-2}]}. \tag{3.20}$$

In this we now replace the unknown parameters μ_θ and σ_θ^2 by their estimates as given above. In extreme cases the estimates reduce to the overall mean proportion or to R_1/m.

In an alternative and equivalent version (3.20) is the mean of a posterior distribution.

A second argument uses the beta distribution. It is easily shown that the posterior distribution of θ_1 is also of the beta form with parameters $(r_1 + \alpha, m - r_1 + \beta)$. If we replace α and β by their moment-based estimates we recover the previous solution; it will, however, be more efficient to use the maximum likelihood estimates of α and β.

In applying empirical Bayes methods it is important to explain as much of the variation as possible systematically before applying empirical Bayes ideas to the remaining random variation.

Example 3.3 Association between smoking and lung cancer (continued)
Mariotto (1988) has made an empirical Bayes analysis of the data of Table 2.9 summarizing 14 studies. From the sth study an estimate $\tilde{\Delta}_s^{(w)}$ of the logistic difference Δ_s is obtained. Suppose that interest focuses on individual Δ_s. We shall treat $\tilde{\Delta}_s^{(w)}$ as normally distributed around Δ_s with variance $v_s^{(w)}$; it is immaterial what method of estimation is used, although here we have employed empirical logit transforms.

Suppose now further that $\Delta_1, \ldots, \Delta_{14}$ are independently normally distributed with mean v and variance τ^2, implying that $\tilde{\Delta}_s$ is normal with mean v and variance $\tau^2 + v_s^{(w)}$, where we make the approximation of treating the $v_s^{(w)}$ as known. From this and the natural assumptions of the independence of $\tilde{\Delta}_1, \ldots, \tilde{\Delta}_{14}$ we can write down the log likelihood of $\tilde{\Delta}_1, \ldots, \tilde{\Delta}_{14}$ and thereby obtain maximum likelihood estimates $\hat{v} = 1.69$, $\hat{\tau} = 0.639$.

The empirical Bayes estimate of Δ_s can now be obtained as a weighted average of $\hat{v}$ and $\tilde{\Delta}_s^{(w)}$.

Table 3.4 compares the resulting estimates and their standard errors. The general nature of the empirical Bayes adjustment is clear; initial estimates of low precision are moved most and the adjustment is towards $\hat{v}$. The nominal standard errors of the empirical Bayes

Table 3.4. *Smoking and lung cancer. Direct and empirical Bayes estimates of logistic differences (Mariotto, 1988)*

Study	Direct estimate		Empirical Bayes estimate	
	$\tilde{\Delta}_s^{(w)}$	Standard error	$\tilde{\Delta}_s^{(EB)}$	Standard error
1	1.83	0.653	1.76	0.456
2	1.90	0.607	1.80	0.441
3	1.51	0.463	1.58	0.374
4	0.64	0.327	0.86	0.292
5	1.74	0.212	1.74	0.200
6	2.61	0.370	2.38	0.319
7	0.25	0.546	0.86	0.415
8	2.27	0.401	2.11	0.339
9	1.79	0.625	1.74	0.447
10	1.37	0.268	1.42	0.247
11	3.81	0.524	2.96	0.405
12	1.18	0.273	1.26	0.251
13	1.46	0.173	1.48	0.167
14	1.81	0.493	1.77	0.390

estimates are less than those of the unadjusted estimates.

The standard errors for the empirical Bayes estimates, and indeed the full justification of the shrinkage procedure, assume normality of the prior distribution and moreover ignore errors of estimation for the parameters (v, τ^2); a more refined calculation suggests that the latter effect is negligible.

3.4 Errors of measurement

3.4.1 General remarks

Normally in studies of dependency one is interested in the relation between observed response variables and observed explanatory variables, taking both as directly typified by the data under analysis. Even if some part of the observed haphazard variability is due to measurement error it may not be appropriate to separate this from other components, e.g. from the natural variability present. In other situations, however, it may be fruitful to postulate an unobserved 'true' response and/or an unobserved 'true' vector of explanatory variables for each individual, these to be distinguished from the 'observed' values, and then to consider the relation between the 'true' variables. This may be useful to assess 'what is really happening' or to give some idea of whether the introduction of improved measurement techniques would materially change the conclusions.

3.4.2 Normal theory

Before studying binary response variables it is useful to review what happens with continuous, and in particular normally distributed, response variables. Addition of a random error of measurement of conditional mean zero at all levels of the explanatory variables does not affect the regression relation, but does, of course, inflate the mean square for deviations from regression. Provided that the variance of the measurement error is constant, its estimation from suitably replicated observations is straightforward, and the effect on the primary conclusions readily assessed by isolating the relevant components of variance. Of course, if the additional variance depends on the explanatory variables, the situation is much more complicated. So far as we know, this possibility has been little discussed in either a practical or theoretical context. Of course if the mean error depended on the explanatory variables the regression relation would be changed.

The effect of measurement errors in the explanatory variables is more complicated. Except in rather special cases, a linear regression on 'true' explanatory variables is converted into a non-linear regression on 'observed' explanatory variables, although the distortion of shape is likely to be in most cases small and thus of at most minor practical importance. Usually it is adequate to regard the effect of errors of measurement as an attenuation, i.e. a flattening of the relationship. Further we shall, largely for convenience, regard the explanatory variables as random and having therefore a variance. If there is a single explanatory variable, the relation between the regression coefficients is

$$\beta_{(m)} = \beta_{(t)} \sigma_{(t)}^2 / \sigma_{(m)}^2, \qquad (3.21)$$

where the subscripts m, t refer to measured and true quantities and both σ^2's refer to the variance of the explanatory variable, x. Thus

$$\sigma_{(m)}^2 = \sigma_{(t)}^2 + \sigma_{(e)}^2, \qquad (3.22)$$

where $\sigma_{(e)}^2$ denotes the measurement variance of x. When there is a vector of explanatory variables the corresponding relation is

$$\beta_{(m)} = \Sigma_{(m)}^{-1} \Sigma_{(t)} \beta_{(t)}, \qquad (3.23)$$

where the Σ's are covariance matrices of the explanatory variables and $\Sigma_{(m)} = \Sigma_{(t)} + \Sigma_{(e)}$, the last term being the covariance matrix of the measurement errors.

Now in addition to a general attenuation effect implicit in (3.23) there is the possibility of a change in the apparent order of importance of the explanatory variables arising when important component variables are subject to relatively large measurement errors, when their effect may get 'transferred' to other correlated components with relatively smaller measurement errors. It is worth studying this effect quantitatively in a very simple special case.

Suppose that there are two correlated explanatory variables x_1 and x_2, that Y is related to $x_{1(t)}$, the true value of x_1, but not to the true value of x_2, and that $x_{1(m)}$ but not x_2 is subject to an error of measurement. We can without loss of generality choose the units of the variables so that

$$Y = x_{1(t)} + \eta,$$

with $\text{var}(\eta) = \sigma_\eta^2$, and so that $x_{1(t)}$ and x_2 have unit variance and correlation ρ. Denote the measurement variance of x_1 by $\sigma_{(e)}^2$, so that

$$\text{var}(x_{1(m)}) = 1 + \sigma_{(e)}^2.$$

It is now straightforward to calculate the form of the least squares regression of Y on $x_{1(m)}$ and x_2.

The simplest result is that the expected regression coefficient on x_2 exceeds that on $x_{1(m)}$ as soon as

$$\sigma_{(e)}^2 > 1/\rho - \rho. \qquad (3.24)$$

The comparison can be made in other ways with only slightly different results. Thus if the expected regression coefficients are standardized by their standard errors from 'simple regression', (3.24) is replaced by

$$\sigma_{(e)}^2 > 1/\rho^2 - 1, \qquad (3.25)$$

with a slightly more complicated formula if standard errors from the multiple regression are used. Note that if $\rho = 1 - \varepsilon$, with ε small, (3.24) and (3.25) are replaced respectively by

$$\sigma^2_{(e)} > 2\varepsilon + \varepsilon^2 + O(\varepsilon^3), \quad \sigma^2_{(e)} > 2\varepsilon + 3\varepsilon^2 + O(\varepsilon^3). \tag{3.26}$$

Thus if $\rho = 0.9$, the 'preference' on the average for regression on the 'wrong' explanatory variable, x_2, takes over at $\sigma^2_{(e)} \simeq 0.2$, $\sigma_{(e)} \simeq 0.45$, whereas at $\rho = 0.99$ the corresponding values are $\sigma^2_{(e)} \simeq 0.02$, $\sigma_{(e)} = 0.14$. Thus the effect under discussion is likely to be important, at least for just two explanatory variables, only at quite high values of ρ.

To use the above results, and especially (3.21)–(3.23), information is needed about the magnitude of measurement errors. Such information can be obtained most directly and satisfactorily either from an independent set of data or via the inclusion in the data of suitable internal replication from which to estimate the relevant component of variance. A third possibility is to do a sensitivity analysis in which a series of trial values of $\sigma^2_{(e)}$, or of the elements of the matrix $\Sigma_{(e)}$, is used to assess whether the main conclusions of an analysis ignoring measurement errors are seriously threatened by values of $\sigma^2_{(e)}$ likely to be present.

3.4.3 Errors in a binary response

An error in measuring or recording a binary response is a misclassification, i.e. the recording of 'true' 1 as a 0 or vice versa. Such errors are characterized by two probabilities, π_{10}, and π_{01}, respectively. When explanatory variables are present in a form that makes the probability of success vary over a wide range, it becomes realistic to consider at least two cases.

In the first, the misclassification probabilities

$$\pi_{01} = \text{prob}(Y_{(m)} = 1 \mid Y_{(t)} = 0), \qquad \pi_{10} = \text{prob}(Y_{(m)} = 0 \mid Y_{(t)} = 1)$$

are the same for all individuals, i.e. do not depend on the explanatory variables. Also errors of misclassification are assumed to occur independently for all individuals. The second possibility is that the misclassification probabilities depend on x. Clearly this could occur in many ways; we study just one below.

The situation with constant misclassification probabilities will be analysed first. The general linear logistic model for $Y_{(t)}$

$$\text{prob}(Y_{(t)} = 1) = \frac{e^{x\beta}}{1 + e^{x\beta}}$$

is converted into

$$\text{prob}(Y_{(m)} = 1) = \frac{\pi_{01} + (1 - \pi_{10})e^{x\beta}}{1 + e^{x\beta}}. \tag{3.27}$$

Note that as $x\beta \to \pm \infty$ this tends to non-unit and non-zero asymptotes. Thus, unlike the normal case, the shape of the regression function is changed by the errors in the response variable. If sufficient data are available in both lower and upper 'tails' of the curve, fitting of (3.27) by maximum likelihood is feasible taking $(\beta, \pi_{01}, \pi_{10})$ as unknown parameters. The model is, however, no longer within the exponential family. Alternatively, as noted in Section 3.4.2, independent estimates of π_{01}, π_{10}, may be available.

If all the probabilities remain in the 'central' part of the range, where the logistic function is to a reasonable approximation linear, then (3.27) also is linear: if

$$e^z/(1 + e^z) \simeq \tfrac{1}{2} + kz,$$

then

$$\text{prob}(Y_{(m)} = 1) \simeq \tfrac{1}{2}(\pi_{01} + 1 - \pi_{10}) + k(1 - \pi_{01} - \pi_{10})x\beta,$$

showing an attenuation of the slope by the factor $1 - \pi_{01} - \pi_{10}$.

There are many possible ways in which the misclassification probabilities might depend on the explanatory variables: while in principle this dependence could be represented parametrically in some plausible fashion and the augmented vector of parameters estimated, this is likely to be a feasible approach only with extensive high-quality data.

One very special case arises when the tolerance model of Section 2.1 is directly relevant. If we take the probit form this means that there is an unobserved random variable, Γ, normally distributed with mean $x\gamma$ and unit variance such that $Y = 1$ if and only if $\Gamma > 0$, so that $\text{prob}(Y = 1) = \Phi(x\gamma)$. Now suppose that error of measurement of response can be approximately treated as an error added to Γ. If this error is normally distributed with zero mean the resulting observed response is determined by a random variable of mean $x\gamma$ and variance $\tau^2 = 1 + \sigma_m^2$ leading to a probit relation $\Phi(\tau^{-1}x\gamma)$, i.e. to an attenuation of the regression coefficients by the constant factor τ^{-1}.

Since for most empirical purposes probit and logit regressions are indistinguishable, it follows that in logit regression also this particular kind of error in the response variable is reflected in an attenuation of the regression coefficients.

A full treatment for binary responses of errors of measurement in the explanatory variable is complicated (Stefanski, 1985). Again, however, a simplified analysis can be based on the probit model plus the assumptions that the explanatory variables and their errors of observation have multivariate normal distributions. For simplicity suppose first that there is a single explanatory variable $x_{(t)}$ normally distributed with mean zero and variance $\sigma^2_{x(t)}$ and that a probit model $\text{prob}(Y = 1) = \Phi(\alpha + x_{(t)}\gamma)$ applies corresponding to the latent variable Γ having conditional mean $\alpha + x_{(t)}\gamma$ and unit conditional variances. If now $x_{(m)} = x_{(t)} + x_{(e)}$, where $x_{(e)}$ is a sampling error and obvious independence assumptions hold, then Γ has, conditionally on $x_{1(m)}$, mean and variance

$$\alpha + \gamma\sigma^2_{(t)}/\sigma^2_{(m)}, \qquad 1 + \gamma^2\sigma^2_{(t)}\sigma^2_{(e)}/\sigma^2_{(m)} = \omega^2,$$

say, so that the probit relation for $x_{(t)}$ is converted into

$$\Phi\left(\frac{\alpha}{\omega} + \frac{\sigma^2_{(t)}}{\omega\sigma^2_{(m)}}x_{(m)}\gamma\right),$$

corresponding to a shift in mean and attenuation of slope. There is a corresponding result for multiple regression. Note again that the same result should apply approximately to logistic relations provided we are not working in regions of low or high probability.

Bibliographic notes

The possibility of overdispersion to binomial, Poisson and similar data has long been recognized and was initially dealt with by inflation of standard errors by a suitable factor (Bartlett, 1937; Finney, 1952). Quasi-likelihood (Wedderburn, 1974; McCullagh, 1983; McCullagh and Nelder, 1983; Engel, 1983) is an elaboration of this idea. For the efficiency of quasi-likelihood relative to a fully parametric discussion, see Cox (1983) and, especially, Firth (1987). For quasi-likelihood with non-constant overdispersion, see Kupper, Portier and Hogan (1986). For the extension to stochastic processes, see Godambe and Heyde (1987).

The alternative to quasi-likelihood is fully parametric formulation. For the beta-binomial distribution, see, in particular, Williams (1975, 1982).

Empirical Bayes methods, of a particularly ingenious kind in which the prior distribution is non-parametric, were introduced by Robbins (1956), but in the present discussion we have concentrated on a fully parametric formulation. Cox (1975) gave a rather complicated procedure for adjusting confidence limits for errors of estimation in the parameters of the prior distribution. For a detailed account of empirical Bayes methods, see Maritz (1989).

The study of regression with errors in the explanatory variables has a long history and an extensive literature; see the book of Fuller (1987). Nearly all this work is, however, concerned with normal-theory linear models or fairly direct generalizations thereof. The comparable discussion for logistic models, or more broadly for generalized linear models, is much more difficult (Carroll, Spiegelman, Lan, *et al.*, 1984; Stefanski, 1985; Stefanski and Carroll, 1985, 1987).

Errors of observation in binary responses are modelled through an extended generalized linear framework by Palmgren and Ekholm (1987); see also Palmgren (1987) and for an early discussion Bross (1954).

CHAPTER 4

Some related approaches

4.1 Introduction

In the previous chapters we have placed some emphasis on the distinction between responses and explanatory variables, this distinction dictating the questions judged relevant as well as the kinds of formal representation and analysis to be used. We now discuss studies in which, for important practical reasons connected with convenient collection of data, the roles of response and explanatory variables appear to be reversed. The general idea is widely applicable. The two most important areas of application are probably epidemiology, in the context of case-control studies, and econometrics, in the context of choice-based sampling. In Section 4.4 we discuss the somewhat related issue of the connection between logistic regression and discriminant analysis.

The nature of and motivation for case-control studies can be explained briefly as follows. An observational prospective study of the relation between explanatory variables and the occurrence of a rare condition, or say death from a rare cause, would proceed roughly as follows. For a suitable group of individuals, explanatory variables are measured. The individuals are then followed for an appropriate time and a response measured, for example death within 10 years from a specific cause, or not. The disadvantages of this procedure are as follows.

1. If the condition is rare, a large initial sample will be necessary in order to obtain enough 'cases' to produce reasonable precision and in a sense the very large amount of information collected on 'non-cases' adds little to the precision of the contrast of interest.
2. Such a study takes a long time to complete and as a result is expensive.

A corresponding retrospective study would proceed broadly as follows. A group of individuals (the cases) showing a rare response of interest is identified and a suitable group of control (non-cases) chosen, possibly by matching one or more controls to each case. Then the explanatory variables are collected for each individual, this usually involving questioning an individual or relatives about past behaviour. Then the distribution of explanatory variables for cases is compared with that for controls. The advantage of this procedure is that it is relatively quick and inexpensive. The disadvantages are the difficulties in specifying an appropriate control group and the possibility of serious bias in determining historical values of explanatory variables when the response of interest is known.

A roughly comparable situation in econometrics arises if it is required to compare individuals using, say, different methods of transport to work, interest being focused on one rather rarely used mode. It will then often be sensible to collect data on all available individuals using the rare mode, together with observations on matching controls.

The following brief discussion starts by setting out in Section 4.2 some simple models for prospective studies and then in Section 4.3 examining how these are connected with the analysis of corresponding retrospective studies.

4.2 Prospective studies

Consider a prospective study, for definiteness a randomized experimental design, in which for a group of individuals explanatory variables are measured, each individual is randomly assigned to one of two treatments and subsequently a binary response is recorded.

The following models, entirely in the spirit of the earlier discussion, arise quite naturally. Throughout $\lambda_i = \log\{\theta_i/(1 - \theta_i)\}$ is the logistic transform of the probability of success θ_i for the ith individual. All models include a treatment effect, and $t_i = 1$ if the ith individual receives T_1, $t_i = 0$ if it receives T_0. The row vector of other explanatory variables for the ith individual is denoted by x_i.

1. Simple treatment effect. Here x_i is ignored and we write

$$\lambda_i = \alpha + \Delta t_i, \tag{4.1}$$

where Δ is the treatment effect.

2. Treatment effect plus arbitrary x dependence. Here

$$\lambda_i = \alpha(x_i) + \Delta t_i, \tag{4.2}$$

where α (.) is an arbitrary function. This is relevant when there are relatively few distinct values of x_i, as when a number of factors each with a small number of possible levels are cross-classified. The analysis for Δ is that for the combination of 2×2 tables.

3. Treatment effect plus linear x dependence. Here we specialize (4.2) to

$$\lambda_i = \alpha_0 + x_i\alpha_1 + \Delta t_i. \tag{4.3}$$

This is more natural than (4.2) when the explanatory variables in the row vector x_i are quantitative; mixtures of the cases of 2 and 3 may often be useful.

4. Treatment $\times$ x interaction. Here we allow for the possibility that the treatment effect varies systematically with x. Corresponding respectively to (4.2) and (4.3) we can write

$$\lambda_i = \alpha(x_i) + \Delta_0 t_i + (x_i\Delta_1)t_i, \tag{4.4}$$

$$\lambda_i = \alpha_0 + x_i\alpha_1 + \Delta_0 t_i + x_i\Delta_1 t_i. \tag{4.5}$$

Here it would often be sensible to constrain some of the elements of Δ_1 to be zero, so that interest is focused on components of interaction of particular interest or potential importance.

The analysis of all these models is via the methods set out in earlier chapters.

4.3 Retrospective studies

We now consider the retrospective analyses that correspond to equations (4.1)–(4.5), starting with the case where there is a single binary explanatory variable t. Suppose that initially there is a population of individuals cross-classified by two binary variables T and Y. Denote by θ_{ij} the proportion of the population with $T = i$ and $Y = j$; see Table 4.1(a). Now in a prospective study separate samples are taken from the sub-populations, conditional first on

Table 4.1 *Cross-classification of population by two binary features*

(a) Population				(b) Prospective				(c) Retrospective		
	Y				Y				Y	
	0	1			0	1			0	1
T 0	θ_{00}	θ_{01}	$\theta_{0\cdot}$	T 0	$\theta_{00}/\theta_{0\cdot}$	$\theta_{01}/\theta_{0\cdot}$ 1		T 0	$\theta_{00}/\theta_{\cdot0}$	$\theta_{01}/\theta_{\cdot1}$
1	θ_{10}	θ_{11}	$\theta_{1\cdot}$	1	$\theta_{10}/\theta_{1\cdot}$	$\theta_{11}/\theta_{1\cdot}$ 1		1	$\theta_{10}/\theta_{\cdot0}$	$\theta_{11}/\theta_{\cdot1}$
	$\theta_{\cdot0}$	$\theta_{\cdot1}$							1	1

$T = 0$ and then on $T = 1$, so that the response Y has a distribution depending on the explanatory variable t. The logistic difference defined via the standard model (4.1) is derived via the conditional probabilities in Table 4.1(b) and is

$$\Delta = \log\left(\frac{\theta_{11}/\theta_{1\cdot}}{\theta_{10}/\theta_{1\cdot}}\right) - \log\left(\frac{\theta_{01}/\theta_{0\cdot}}{\theta_{00}/\theta_{0\cdot}}\right) = \log\left(\frac{\theta_{11}\theta_{00}}{\theta_{10}\theta_{01}}\right) \qquad (4.6)$$

For the corresponding retrospective data, provided that controls $(y = 0)$ and cases $(y = 1)$ can be treated as randomly sampled from the conditional distributions given $Y = 0$ and given $Y = 1$ as shown in Table 4.1(c), we have that the logistic difference associated with the formal response T is

$$\log\left(\frac{\theta_{11}/\theta_{\cdot1}}{\theta_{01}/\theta_{\cdot1}}\right) - \log\left(\frac{\theta_{10}/\theta_{\cdot0}}{\theta_{00}/\theta_{\cdot0}}\right) = \Delta, \qquad (4.7)$$

so that the standard analysis of the retrospective 2×2 table yields inference about the same parameter as the prospective study.

This simple conclusion depends critically on the use of the logistic difference to characterize the effect of interest. If, for instance, interest lay in a difference between 'groups' in the probabilities of 'success' $(Y = 1)$, i.e. in

$$\delta = \theta_{11}/\theta_{1\cdot} - \theta_{01}/\theta_{0\cdot} = \theta_{11}/(\theta_{10} + \theta_{11}) - \theta_{01}/(\theta_{00} + \theta_{01}), \quad (4.8)$$

it is clear that this can be estimated directly from the prospective study, but cannot be estimated directly from the retrospective study. This is essentially because the retrospective study gives no information about the marginal probability of Y. To estimate δ some additional information has to be available. One possibility is to have an independent estimate of the marginal probability that $Y = 1$. If from the retrospective study there m_0, m_1 controls ($y = 0$) and cases ($y = 1$), with s_{10}, s_{11} showing a positive value of the explanatory variable ($t = 1$), and if in the independent study there are n' individuals, r' of whom are cases ($y = 1$), then there are essentially three observed proportions to be matched with three independent θ's. This leads to

$$\hat{\delta} = \frac{r'}{n'} \left\{ \frac{s_{11}/m_1}{s_{10}/m_0 + (r'/n')(s_{10}/m_0 - s_{11}/m_1)} \right. $$
$$\left. - \frac{(1 - s_{11}/m_1)}{(1 - s_{10}/m_0) + (r'/n_1')(s_{10}/m_0 - s_{11}/m_1)} \right\}.$$

Another possible situation is to have data on independent prospective and retrospective studies on the same population. The mutual consistency of the studies can be tested and, assuming consistency, any required combination of parameters estimated.

For the remainder of the discussion, we concentrate on logistic parameters. Discussion under (4.2) is a direct extension of that for a single 2×2 table. Provided that at each level of the explanatory variable cases and controls are randomly sampled from the relevant sub-populations, we have a series of 2×2 tables with a common logistic difference, and analysis proceeds in principle via conditional maximum likelihood. A particular case is that of the matched case-control study in which, for each case, one control individual (or sometimes a small number k of control individuals) is chosen from the control sub-population having the same values of the primary explanatory variables as the case. We are then faced with a large number of 2×2 tables with a common logistic difference Δ and with column sums $(1, 1)$ or more generally $(1, k)$. If each table is labelled with a vector of explanatory variables x, we can fit a model (4.4) allowing for treatment $\times$ x interaction.

A quite general discussion under the models (4.2)–(4.5) can be given as follows. Suppose first that the underlying 'causal' model is the prospective one, say (4.2). Now suppose that individuals are assigned to 'treatments' by some mechanism such that the probability that an individual is assigned to $T = 1$, say, depends in an unknown way on x

$$\text{prob}(T_i = 1 | x_i) = \pi(x_i).$$

Once this assignment is made, the resulting values of the response Y_i

are determined via (4.2). This generates a population of values of (T_i, Y_i) which can be sampled retrospectively. The relevant probabilities in this retrospective sampling are

$$\text{prob}(T_i = 1 \mid Y_i = 1, x_i) = \text{prob}(T_i = 1 \text{ and } Y_i = 1 \mid x_i)/\text{prob}(Y_i = 1 \mid x_i)$$

$$= \frac{\pi(x_i)\exp[\alpha(x_i) + \Delta]/\{1 + \exp[\alpha(x_i) + \Delta]\}}{\dfrac{\pi(x_i)\exp[\alpha(x_i) + \Delta]}{1 + \exp[\alpha(x_i) + \Delta]} + \dfrac{[1 - \pi(x_i)]\exp[\alpha(x_i)]}{1 + \exp[\alpha(x_i)]}}$$

with a similar expression for $\text{prob}(T_i = 0 \mid Y_i = 1, x_i)$, etc. It follows that

$$\log \frac{\text{prob}(T_i = 1 \mid Y = y_i, x_i)}{\text{prob}(T_i = 0 \mid Y = y_i, x_i)} = \Delta y_i + \alpha^*(x_i), \qquad (4.9)$$

where $\alpha^*(x_i)$ depends on $\alpha(x_i)$ and on $\pi(x_i)$. If an interaction term is included there is a direct modification of (4.9).

The conclusion is then that we may fit a logistic regression to the retrospective data and that the coefficient of y_i has a direct prospective interpretation, but that the terms involving x_i depend on analogous prospective terms and on unknown probabilities of 'treatment' allocation and therefore in general cannot be usefully interpreted.

Example 4.1 Peptic ulcers and blood group
In a retrospective study of the possible effect of blood group on the incidence of peptic ulcers, Woolf (1955) obtained data from three cities. Table 4.2 gives for each city data for blood groups O and A only. In each city, blood group is recorded for peptic-ulcer subjects and for a control series of individuals not having peptic ulcer.

We want really to study how the probability of peptic ulcer depends on blood group, i.e. to use occurrence of peptic ulcer as a response and blood group as an explanatory variable. For fairly obvious reasons,

Table 4.2 *Blood groups for peptic ulcer and control subjects*

	Peptic ulcer		Control	
	Group O	Group A	Group O	Group A
London	911	579	4578	4219
Manchester	361	246	4532	3775
Newcastle	396	219	6598	5261

it is convenient to collect data in an inverse fashion, taking a set of peptic-ulcer patients and a set chosen from non-peptic-ulcer individuals and then observing blood group for each individual. The possibility of using data of this type to answer the question of interest depends on the equivalence of (4.6) and (4.7).

Within city s ($s = 1, 2, 3$) we assume a logistic model such that

$$\lambda_s = \alpha_s + \Delta t,$$

where $t = 1$ for peptic ulcer subjects and $t = 0$ for control subjects. The maximum likelihood estimate $\hat{\Delta} = -0.330$, with standard error 0.042. The assumption of a common parameter Δ for each city is confirmed by the likelihood ratio goodness-of-fit statistic, $\chi^2 = 2.97$, with two degrees of freedom.

Now if the probability of having an ulcer is small, then

$$\Delta = \log \left\{ \frac{\text{prob}(\text{ulcer}|A)}{\text{prob}(\text{ulcer}|O)} \cdot \frac{\text{prob}(\text{no ulcer}|O)}{\text{prob}(\text{no ulcer}|A)} \right\}$$

$$\simeq \log \left\{ \frac{\text{prob}(\text{ulcer})|A)}{\text{prob}(\text{ulcer})|O} \right\},$$

so that $e^{\hat{\Delta}} = 0.719$ estimates the ratio of the probability of having an ulcer in blood group A to that of having one in blood group O; a 95 per cent confidence interval for the ratio is (0.663, 0.780).

An analysis, as given for this example by Cox and Snell (1981, p. 155), based on the empirical logistic transform (and similar to that for Example 2.13), leads to the same numerical estimates and goodness-of-fit statistic as above. This is to be expected with the large sample sizes.

4.4 Relation between discriminant analysis and logistic regression

We now consider the relation between two intimately related and yet conceptually quite different techniques, namely discriminant analysis and logistic regression. We consider first the corresponding general formulations and then the more specific details of the techniques of analysis involved.

The general framework for logistic regression, at least in prospective contexts, should be clear from the earlier chapters. That is, for each individual there is a row vector, x, of explanatory variables and for any given x in the relevant range, it is reasonable to think of a well-defined probability that the binary response, y, is one. This in effect presupposes a stable statistical relation such that in the context in question once x is given a probability is determined. The distribution of x is not directly relevant to the definition. Usually y is in some sense (partially) determined by x and ideally the vector x would be complete enough virtually to determine y exactly, although of course we often have to be content with much less than that. The linear logistic regression is a particular specification of this relation and is conceptually just a special case of a regression or dependence relation.

In discriminant analysis the primary notion is that there are two distinct populations, defined by $y = 0, 1$, usually two intrinsically different groups, like two species of bacteria or plants, two different kinds of product, two distinct but rather similar drugs, and so on. Within each of these populations there is a distribution of a set of properties, x. That is there are two probability densities $f_0(x)$ and $f_1(x)$. Now the emphasis is on these densities, which are supposed to be stable well-defined characteristics of the two populations.

Essentially the focus in discriminant analysis is on the question: how do the two distributions differ most sharply? Often this is put in more specific form as follows. There is given a new vector x' from an individual of unknown y. What can we say about that y, i.e. about the population from which the individual was drawn? There are various elaborations of this idea, including the possibility that the individual came from neither of the two base populations. Note that the emphasis is strongly on the distributions of x within the two populations.

There are now two rather different situations for consideration. In the first the relative frequencies with which the two populations generate data are not defined, i.e. may change relatively arbitrarily under hypothetical repetition. There is then no sense in which we can represent y by a random variable, i.e. we cannot consider its probability distribution either marginally or conditionally on x. Thus logistic regression is not applicable. Various considerations indicate that the statistic for assessing x' is the log likelihood ratio

$$\log f_1(x') - \log f_0(x').\qquad(4.10)$$

Large positive values point to $y = 1$, large negative values to $y = 0$.

Now if the two densities come from the same exponential family with canonical statistic x and with two different parameter values, then it is easily seen that (4.10) is a linear function of the components of x; see Appendix 1. We call the resulting function a linear discriminant function. The most important special case is when f_0 and f_1 are multivariate normal densities with the same covariance matrix Σ and means μ_0 and μ_1. Then (4.10) becomes

$$-\tfrac{1}{2}(\mu_1 \Sigma^{-1} \mu_1^T - \mu_0 \Sigma^{-1} \mu_0^T) + x\Sigma^{-1}(\mu_1 - \mu_0)^T, \qquad (4.11)$$

which is the population linear discriminant function.

The second situation, still within the broad framework of discriminant analysis, is that there are physically defined probabilities π_0, π_1 that y is 0, 1 with $\pi_0 + \pi_1 = 1$, so that we can represent the population for an arbitrary individual by a random variable Y and the full properties of such an individual by a vector random variable (Y, X); the functions $f_0(x)$ and $f_1(x)$ now specify conditional densities of X given $Y = 0, 1$. For the new individual of known x' but unknown Y, we have by Bayes's theorem that

$$\text{prob}(Y = 1 | X = x') = \frac{f_1(x')\pi_1}{f_0(x')\pi_0 + f_1(x')\pi_1}$$

so that

$$\log\{\text{prob}(Y = 1 | X = x')/\text{prob}(Y = 0 | X = x')\}$$
$$= \log(\pi_1/\pi_0) + \log\{f_1(x')/f_0(x')\}, \qquad (4.12)$$

defining a logistic regression equation in which the prior probabilities are isolated into a single term. When we have a linear discriminant function in the sense mentioned above there results a linear logistic regression. This happens when, but by no means only when, the conditional distributions of X are normal with the same covariance matrix.

In this formulation the logistic regression equation emerges as a natural way of describing the uncertainty in attaching a value of y to a new individual with a known value of x; alternatively, if that particular problem is somewhat notional, we can regard the right-hand side of (4.12) as a function of x' describing the 'criterion' in terms of which the two distributions of X most differ.

Kay and Little (1987) have used the general relation (4.12) in the context of logistic regression to motivate ways of testing goodness of fit of the simple linear form. Thus if it is reasonable to think of multivariate normal distributions with unequal covariance matrices, quadratic terms in x' become involved, whereas if gamma distributions are suspected inclusion of log components of x' is indicated.

So far we have discussed only the formal structure underlying the two approaches. We now compare the methods of analysis. One approach to linear discriminant analysis which in a sense is purely descriptive is to look for that linear combination of the components of x which most strongly separates the two groups, usually in the sense of maximizing the square of the difference between the group means divided by its formal variance, the latter estimated assuming a common but arbitrary covariance matrix for the components of x. Note that in this approach the discriminant function is indeterminate to a constant multiplier, at least until some normalizing condition is imposed.

It is a remarkable consequence of the geometry of the estimation problem that not only is this technique equivalent to the formal *linear* regression of the binary variable y on the vector x, treated as fixed, but that under normal-theory assumptions exact tests of regression coefficients are obtained by pretending that the 'fixed' binary y is normal and that the random multivariate normal x is fixed.

Suppose, initially, however, that we take the second formulation above with parameters unknown. With only slight loss of generality we take the two conditional densities to be of common parametric form with some common parameters; that is, we write

$$f_0(x) = g(\psi, \lambda_0), \quad f_1(x) = g(\psi, \lambda_1),$$

for some known function g. The unknown parameters are thus ψ, λ_0, λ_1 and, say, π_0. The multivariate normal case with the common covariance matrix is clearly included. We assume available n_0, n_1 individuals from the two groups, each individual having its full x vector of observations; we assume all individuals are independent and generated by the full probability model we have described.

We now apply the method of maximum likelihood to the estimation of unknown parameters. Now for the ith individual we can write the contribution to the likelihood either as

$$\text{prob}(Y = y_i)f_Y(x_i) \qquad \text{for } y_i = 0, 1, \tag{4.13}$$

or as

$$\text{prob}(Y = y_i | X = x_i)f_X(x_i). \tag{4.14}$$

If we apply maximum likelihood to the first version, (4.13), it is easy to see that $\hat{\pi}_0 = n_0/(n_0 + n_1)$ and that the remaining parameters are estimated as from two independent samples of sizes n_0 and n_1 (from the densities f_0 and f_1). In the special case of the multivariate normal densities of the equal covariance matrix this leads to the substitution into the population form of the standard estimates of the mean and covariance matrix. As noted above this is closely related to the result of least squares linear regression of the binary y on x.

If we apply maximum likelihood to the second version (4.14) of course the same answer results. Note, however, that if we used only the first factor of (4.14) we would be using techniques of logistic regression analysis, linear regression in simple cases, but in general using the form of regression equation implied by the special assumptions made. It follows that the logistic regression analysis is in general inefficient in that the second factor of (4.14) does contain information about relevant parameters and this information has been totally discarded. Efron (1975) and Ruiz (1989) have investigated this loss, the former using discriminant misclassification rates as a criterion and the latter estimating efficiency. Up to one-third of loss of efficiency can occur. Note, however, that this information is based on the assumed functional form of the conditional distributions of x in the two populations and in many applications these assumptions would be unreasonable.

Under the assumption of multivariate normal densities with common covariance matrix Σ the estimate $\hat{\beta}_{(d)}$ of coefficients determined by a discriminant approach is directly proportional to $\tilde{\beta}_{(l)}$ obtained by substitution of the maximum likelihood estimates $\hat{\Sigma}$, $\hat{\mu}_0$ and $\hat{\mu}_1$ into the population discriminant (4.11). Thus, from (4.11),

$$\tilde{\beta}_{(l)} = \hat{\Sigma}^{-1}(\bar{x}_1 - \bar{x}_0)^{\mathrm{T}}.$$

The estimate $\hat{\beta}_{(d)}$ is given by solution of the least squares equations

$$S\hat{\beta}_{(d)} = (n_0 n_1/n)(\bar{x}_1 - \bar{x}_0)^{\mathrm{T}},$$

where S is the matrix of total sums of squares and products and n_0, n_1 $(n = n_0 + n_1)$ are the numbers responding to $Y = 0, 1$. Since

$$S = n\hat{\Sigma} + (n_0 n_1/n)(\bar{x}_1 - \bar{x}_0)^{\mathrm{T}}(\bar{x}_1 - \bar{x}_0),$$

we have that

$$\hat{\beta}_{(d)} = (n_0 n_1/n)\{1 - (\bar{x}_1 - \bar{x}_0)\hat{\beta}_{(d)}\}\hat{\beta}_{(l)} = k\tilde{\beta}_{(l)},$$

where k is equal to the difference between the total sum of squares

$n_0 n_1/n$ and the sum of squares due to regression. Hence

$$\hat{\beta}_{(d)} = \tilde{\beta}_{(l)} \text{ss}_{\text{res}}/n \simeq \hat{\beta}_{(l)} \text{ss}_{\text{res}}/n, \tag{4.15}$$

where $\hat{\beta}_{(l)}$ is the maximum likelihood logistic regression estimate and ss_{res} is the residual sum of squares obtained when y is regressed on x.

In summary, therefore, the key issue in deciding which viewpoint is the more relevant is the meaningful stability of the conditional probability of Y given x and of the distributions of x within the two sub-populations. If both approaches are applicable, and the logistic regression is effectively linear, logistic regression assumes less, in that given the linear regression the forms of the distributions of x within sub-populations are irrelevant. On the other hand, if multivariate normality or some other specific distributional form can be taken, then the discriminant approach is more efficient.

Example 4.2 Nodal involvement in cancer patients
Table 4.3 (Brown, 1980) gives data for 53 patients receiving surgical treatment for cancer of the prostate. A critical question in determining treatment for patients is whether the cancer has spread to the neighbouring lymph nodes and whether this can be predicted from variables observed before surgery, in particular X-ray reading, x, stage of tumour assessed by palpation, s, grade of tumour as determined by biopsy, g, age of patient at diagnosis and level of serum acid phosphatase. Variables x, s and g are binary $(0, 1)$ with a value 1 denoting the more serious state. Nodal involvement, determined at surgery, is denoted by $Y = 1$; otherwise $Y = 0$.

Logistic regression is a natural approach for this problem since it requires no assumption about the distribution of the explanatory variables, three of which are binary and two continuous; also $\text{prob}(Y = 1)$, conditional on given values of the explanatory variables, is likely to be stable and thus is a meaningful quantity. For comparison, however, we examine both the logistic and discriminant approach.

We transform the acid measurement, taking $z = \log(\text{acid})$, partly because of some obvious skewness in the data and partly because the more symmetrical, or approximately normal, the distribution can be made the closer the agreement between the two approaches is likely to be. Linearity of the logistic regression is also crucial.

The variables which are significant in the logistic regression are x, s, g, $s \times g$ and z. The estimated coefficients are given in Table 4.4.

Table 4.3 *Nodal involvement, Y, for 53 patients with prostate cancer*

Patient	x	s	g	Age	Acid	Y
1	0	0	0	66	48	0
2	0	0	0	68	56	0
3	0	0	0	66	50	0
4	0	0	0	56	52	0
5	0	0	0	58	50	0
6	0	0	0	60	49	0
7	1	0	0	65	46	0
8	1	0	0	60	62	0
9	0	0	1	50	56	1
10	1	0	0	49	55	0
11	0	0	0	61	62	0
12	0	0	0	58	71	0
13	0	0	0	51	65	0
14	1	0	1	67	67	1
15	0	0	1	67	47	0
16	0	0	0	51	49	0
17	0	0	1	56	50	0
18	0	0	0	60	78	0
19	0	0	0	52	83	0
20	0	0	0	56	98	0
21	0	0	0	67	52	0
22	0	0	0	63	75	0
23	0	0	1	59	99	1
24	0	0	0	64	187	0
25	1	0	0	61	136	1
26	0	0	0	56	82	1
27	0	1	1	64	40	0
28	0	1	0	61	50	0
29	0	1	1	64	50	0
30	0	1	0	63	40	0
31	0	1	1	52	55	0
32	0	1	1	66	59	0
33	1	1	0	58	48	1
34	1	1	1	57	51	1
35	0	1	0	65	49	1
36	0	1	1	65	48	0
37	1	1	1	59	63	0
38	0	1	0	61	102	0
39	0	1	0	53	76	0
40	0	1	0	67	95	0
41	0	1	1	53	66	0
42	1	1	1	65	84	1
43	1	1	1	50	81	1

(Contd.)

Table 4.3 (*contd.*)

Patient	x	s	g	Age	Acid	Y
44	1	1	1	60	76	1
45	0	1	1	45	70	1
46	1	1	1	56	78	1
47	0	1	0	46	70	1
48	0	1	0	67	67	1
49	0	1	0	63	82	1
50	0	1	1	57	67	1
51	1	1	0	51	72	1
52	1	1	0	64	89	1
53	1	1	1	68	126	1

Table 4.4 *Comparison of estimated logistic regression coefficients, $\hat{\beta}_{(l)}$ and discriminant analysis coefficients, $\hat{\beta}_{(d)}$*

Variable	$\hat{\beta}_{(l)}$	$\hat{\beta}_{(d)}$	$\hat{\beta}_{(l)}/\hat{\beta}_{(d)}$	$\hat{\beta}_{(l)}/\text{s.e.}\ \hat{\beta}_{(l)}$	$\hat{\beta}_{(d)}/\text{s.e.}\ \hat{\beta}_{(d)}$
x	2.50	0.396	6.3	2.5	3.2
s	3.43	0.460	7.5	2.8	3.4
g	3.97	0.534	7.4	2.6	2.8
$s \times g$	-4.60	-0.656	7.0	2.5	2.7
z	2.84	0.412	6.9	2.2	2.4

For comparison with the linear discriminant estimates we regress Y on these same variables. Inclusion of the interaction $s \times g$ is justified by regarding s, g and $s \times g$ as three variables which define the 2×2 categories formed by s and g.

The relative magnitudes of the logistic and discriminant coefficients are very similar, with the logistic estimates being approximately seven times the discriminant estimates (see Table 4.4). This agrees surprisingly well with a ratio 7.8 obtained from (4.15), which is based upon the assumption of multivariate normality, an assumption which clearly does not hold in this instance.

We note, that as would be expected, the discriminant estimates are notionally more efficient (as measured by the ratio of estimate to standard error). The estimates of precision from the discriminant analysis, however, are based as above on the manifestly false supposition of multivariate normality.

4.5 Some different theoretical formulations

4.5.1 A simple example

In this section we outline the relations between a number of different theoretical formulations, showing in fact that essentially the same analysis can be derived from a number of rather different points of view. This is of theoretical interest and also has the more immediate practical consequence that certain computer programs are, with due care, interchangeable.

We start with the simplest special case. Consider n independent binary trials with constant probability θ of success; let $\lambda = \log\{\theta/(1-\theta)\}$. The individual trials are represented by random variables $Y_1, \ldots, Y_n$; for many purposes one would use the aggregated form in which we summarize the outcomes via the frequencies $n - R$ and R of 0s and 1s. Here $R = \sum Y_j$ has the binomial distribution

$$\text{prob}(R = r) = \binom{n}{r}(1-\theta)^{n-r}\theta^r.$$

The likelihood from either individual or aggregated form is

$$(1-\theta)^{n-r}\theta^r, \tag{4.16}$$

where in the second case there is an additional combinatorial factor.

Next consider two Poisson-based models, one individual and one aggregated. In the individual form we have $2n$ independent Poisson distributed random variables

$$\{Z_{10}, Z_{11}; Z_{20}, Z_{21}; \ldots; Z_{n0}, Z_{n1}\}$$

with means

$$\{\mu_{10}, \mu_{11}; \mu_{20}, \mu_{21}; \ldots; \mu_{n0}, \mu_{n1}\}.$$

In the aggregated form we have just two independent Poisson random variables (Z_0, Z_1) with means (μ_0, μ_1). In both cases the corresponding likelihoods are formed as a product of Poisson contributions.

Now in the former model the contribution from (Z_{j0}, Z_{j1}) can be written

$$\frac{e^{-\mu_{j0}}\mu_{j0}^{z_{j0}}}{z_{j0}!} \frac{e^{-\mu_{j1}}\mu_{j1}^{z_{j1}}}{z_{j1}!} = \frac{e^{-\mu_{j\cdot}}\mu_{j\cdot}^{z_{j\cdot}}}{z_{j\cdot}!} \times \binom{z_{j\cdot}}{z_{j0}}\left(\frac{\mu_{j0}}{\mu_{j\cdot}}\right)^{z_{j0}}\left(\frac{\mu_{j1}}{\mu_{j\cdot}}\right)^{z_{j1}}, \tag{4.17}$$

where $z_{j\cdot} = z_{j0} + z_{j1}, \mu_{j\cdot} = \mu_{j0} + \mu_{j1}$ and the two factors represent respectively the marginal distribution of Z_j, and the conditional distribution given $Z_j = z_j.$.

We now take the special model

$$\log \mu_{j0} = \alpha_j, \qquad \log \mu_{j1} = \alpha_j + \lambda, \qquad (4.18)$$

$$\mu_{j0}/\mu_{j.} = 1/(1 + e^\lambda), \qquad \mu_{j1}/\mu_{j.} = e^\lambda/(1 + e^\lambda) = \theta. \qquad (4.19)$$

It is convenient to express (4.18) not in terms of α_j and λ but in terms of $\eta_j = \mu_{j}$ and λ. Then if we consider the special case in which each pair (Z_{j0}, Z_{j1}) sum to one, we have on multiplying the contributions (4.17) that the likelihood can be written

$$g(\eta_1, \ldots, \eta_n)(1 - \theta)^{n-r}\theta^r, \qquad (4.20)$$

where $r = \Sigma z_{j1}$, $n - r = \Sigma z_{j0}$.

Virtually the same conclusion can be reached from the aggregated form. Here we start, instead of with $(n - R, R)$, by considering two independent Poisson variables (Z_0, Z_1) with means (μ_0, μ_1). We write

$$\log \mu_0 = \alpha, \qquad \log \mu_1 = \alpha + \lambda \qquad (4.21)$$

and take as new parameters $\eta = \mu_0 + \mu_1$ and λ. Then the likelihood for the aggregated Poisson model can be written

$$\frac{e^{-\mu_0}\mu_0^{z_0}}{z_0!} \cdot \frac{e^{-\mu_1}\mu_1^{z_1}}{z_1!} = \frac{e^{-\eta}\eta^{z.}}{z.!}\binom{z.}{z_1}(1 - \theta)^{z_0}\theta^{z_1}; \qquad (4.22)$$

with the identification $z. = n$, $z_1 = r$, the second factor of (4.22) differs from (4.16) by at most the combinatorial factor.

There is a further model closely associated with the 'individual' Poisson model. If, starting from the product of terms in (4.17), we condition on $\Sigma z_j = z..$, we can write the full likelihood as

$$\frac{e^{-\mu..}\mu_{..}^{z..}}{z..!} \cdot \frac{z..!}{\prod z_{0j}!\prod z_{1j}}\prod\left(\frac{\mu_{j0}}{\mu..}\right)^{z_{j0}}\Sigma\left(\frac{\mu_{j1}}{\mu..}\right)^{z_{j1}}, \qquad (4.23)$$

where the second factor is the multinomial distribution corresponding to distributing $z..$ trials among $2n$ cells with probabilities $\phi_{j0} = \mu_{j0}/\mu..$, $\phi_{j1} = \mu_{j1}/\mu..$. If now we take the log linear form

$$\log \phi_{j0} = \varepsilon + \alpha_j, \qquad \log \phi_{j1} = \varepsilon + \alpha_j + \lambda, \qquad (4.24)$$

where it is convenient to insert a normalizing constant ε chosen to make the separate probabilities add to one, then the combined likelihood factorizes into a function of $\varepsilon, \alpha_1, \ldots, \alpha_n$ multiplied by (4.16).

We call (4.24) a log linear multinomial model. There is a close link with (4.18).

Note that so far as the data are concerned we are in the individual Poisson model replacing $y_j = 0$, failure on the jth trial, by two observations $z_{j0} = 1$, $z_{j1} = 0$ and replacing success on the jth trial by two observations $z_{j0} = 0$, $z_{j1} = 1$. Similarly in the aggregated Poisson model we are replacing the observation of r successes in n trials by two notionally independent observations $z_0 = n - r$, $z_1 = r$. Finally in the multinomial model we are in effect regarding the labelling $1, \ldots, n$ also as random, there being n independent trials each in principle capable of any of $2n$ possible outcomes.

The statistical consequences of all this are as follows.

1. If in the more complex models λ, or equivalently θ, is the parameter of interest, the other parameters being nuisance parameters, then inference is made conditionally and we are led to the binomial distribution, i.e. to the same inference as in the initial model. Thus for this particular purpose all the models are equivalent.

2. It is clear immediately from the factorizations (4.20) and (4.22) that, again provided that θ is the parameter of interest, maximum likelihood estimates $\hat{\theta}$ and their asymptotic standard errors calculated via the observed information matrix are the same for the various models. In general if the likelihood is formed from a product $g(\eta)h(\theta)$ of a function of nuisance parameters η and a parameter of interest θ, the log likelihood is the sum of two terms, the maximum likelihood estimate is defined by $h'(\hat{\theta}) = 0$ and the matrix of observed and expected second derivatives has zero mixed (θ, η) elements. Thus, provided the usual asymptotic theory of maximum likelihood estimates holds, the properties of $\hat{\theta}$ depend only on the factor $h(\theta)$.

3. Of course if questions arise other than inference about θ, taking the model as given, then different answers may well arise from the different models. Also examining the adequacy of the model could require different procedures for the different models.

4. The physical relevance of the different models depends on the context. For example, if we were comparing the rates of two independent Poisson processes observed over a pre-assigned time period, the Poisson model would be very natural. Nevertheless for most of the problems discussed in the present book, the direct use of a binary response leading to (4.16) seems the most suitable approach. The modelling of irrelevant features via unknown parameters that

have then to be 'conditioned out' is in most circumstances a tortuous approach. There is, of course, no such thing as a chance happening in probability theory and the identity of these analyses is no accident; it does, however, have to be viewed as both a considerable convenience and as a source of potential conceptual confusion.

4.5.2 Comparison of two binomial variables

We now consider the comparison of two treatments, a problem that has recurred so often above. The binomial model used previously has, for treatment T_0, n_0 trials with R_0 successes and probability of success θ_0, with n_1, R_1 and θ_1 referring similarly to treatment T_1. The likelihood is thus the product of two binomial terms. If, as in the previous discussion we write $\lambda_s = \log\{\theta_s/(1 - \theta_s)\}$, $\lambda_0 = \gamma$, $\lambda_1 = \gamma + \Delta$, we base inference about Δ as in Section 2.1.3 on the factorization of the combined likelihood as

$$\text{prob}(R_0 + R_1 = r.)\,\text{prob}(R_1 = r_1 | R_0 + R_1 = r.), \qquad (4.25)$$

where the second factor is the generalized hypergeometric distribution depending only on Δ and therefore the basis of inference about Δ. Note, however, that

$$\text{prob}(R_0 + R_1 = r.) = \sum_t \text{prob}(R_0 = t)\,\text{prob}(R_1 = r. - t)$$

$$= \sum_t \binom{n_0}{t} \frac{e^{(n_0 - t)\gamma}}{(1 + e^{\gamma)n_0}} \binom{n_1}{r. - t} \frac{e^{(n_1 - r. + t)(\gamma + \Delta)}}{(1 + e^{\gamma + \Delta})^{n_1}} \quad (4.26)$$

and is a complicated function of both γ and Δ. The type of simple factorization which provided the basis of Section 4.5.1 is not achievable by reparameterization: in particular, as we have seen in Section 2.3, the conditional maximum likelihood estimate of Δ is not the same as that obtained by maximizing the whole likelihood, as is fairly clear from the occurrence of Δ in (4.26).

It would be possible to introduce an individual Poisson-based model analogous to that of Section 4.5, but we shall concentrate on an aggregate model. In this the two random variables R_0 and R_1 are replaced by four random variables, namely $(Z_{00}, Z_{01}; Z_{10}, Z_{11})$, having independent Poisson distributions with means $(\mu_{00}, \mu_{01}; \mu_{10}, \mu_{11})$. We now in effect apply (4.21) and (4.22) twice. Write

$$\mu_{0.} = \mu_{00} + \mu_{01}, \quad \theta_0 = \mu_{01}/\mu_{0.}, \quad \mu_{1.} = \mu_{10} + \mu_{11}, \quad \theta_1 = \mu_{11}/\mu_{1.}.$$

Then the likelihood formed from the Poisson terms can be written

$$\frac{e^{-\mu_{0\cdot}}\mu_{0\cdot}^{z_{0\cdot}}}{z_{0\cdot}!} \frac{e^{-\mu_{1\cdot}}\mu_{1\cdot}^{z_{1\cdot}}}{z_{1\cdot}!} \binom{z_{0\cdot}}{z_{00}} \theta_0^{z_{01}}(1-\theta_0)^{z_{00}} \binom{z_{1\cdot}}{z_{10}} \theta_1^{z_{11}}(1-\theta_1)^{z_{10}}$$

$$(4.27)$$

and again, provided that interest is concentrated on θ_0 and θ_1, we consider only the final factor, which is identical to the product of two binomial terms used in the earlier discussion. The log linear form is

$$\log\mu_{00} = \alpha_0, \qquad \log\mu_{01} = \alpha_0 + \lambda_0, \qquad \log\mu_{10} = \alpha_1,$$
$$\log\mu_{11} = \alpha_1 + \lambda_1 = \alpha_1 + \lambda_0 + \Delta. \tag{4.28}$$

The final factor in (4.27) is formed by conditioning on $Z_{0\cdot}, Z_{1\cdot}$. The corresponding multinomial model requires conditioning only on $Z_{\cdot\cdot} = Z_{0\cdot} + Z_{1\cdot}$. With $\mu_{\cdot\cdot} = \mu_{0\cdot} + \mu_{1\cdot}$, we obtain instead of (4.27)

$$\frac{e^{-\mu_{\cdot\cdot}}\mu_{\cdot\cdot}^{z_{\cdot\cdot}}}{z_{\cdot\cdot}!} \frac{z_{\cdot\cdot}!}{z_{00}!z_{01}!z_{10}!z_{11}!} \phi_{00}^{z_{00}} \phi_{01}^{z_{01}} \phi_{10}^{z_{10}} \phi_{11}^{z_{11}}, \tag{4.29}$$

where $\phi_{ij} = \mu_{ij}/\mu_{\cdot\cdot}$.

If we consider the log linear representation

$$\log\phi_{00} = \varepsilon + \alpha_0, \quad \log\phi_{01} = \varepsilon + \alpha_0 + \lambda,$$
$$\log\phi_{10} = \varepsilon + \alpha_1, \quad \log\phi_{11} = \varepsilon + \alpha_1 + \lambda + \Delta, \tag{4.30}$$

subject to the constraint that the ϕ_{ij} sum to one, so that

$$e^\varepsilon\{e^{\alpha_0}(1 + e^\lambda) + e^{\alpha_1}(1 + e^{\lambda+\Delta})\} = 1,$$

then a direct substitution in (4.29) shows that the dependence on λ, Δ is equivalent to (4.26) and that on Δ is specified by the generalized hypergeometric distribution.

The multinomial model is the natural one if we wish to deal with the rows and columns of the 2×2 contingency tables symmetrically. This is right if rows represent not 'treatments' but another response variable; for instance, rows and columns may correspond to presence and absence of two different symptoms. Then $\{\phi_{00}, \phi_{01}, \phi_{10}, \phi_{11}\}$ specifies the distribution of the combined response variable over its four possible outcomes. The essential conclusion is that if association between the two component variables is measured by $\psi = (\phi_{00}\phi_{11})/(\phi_{01}\phi_{10})$, this being 1 for independence, then inference about ψ is carried out by the procedures that we have developed above in a different context.

We have in this section regarded the two independent binomial model as the key one; this is solely for the empirical reason that it seems the most widely appropriate model for the kind of problem we have in mind. In this route the generalized hypergeometric distribution arises by further conditioning justified by theoretical considerations. The other models are in some senses more complicated than the binomial model but lead on suitable conditioning to it.

A final question concerns whether the generalized hypergeometric distribution is ever a natural model on its own right, i.e. whether it is ever natural to start with both margins of the contingency table fixed. In fact the null (central) hypergeometric distribution arises in the randomization theory of experimental design. In the randomization theory of an experiment to compare two treatments, T_0 and T_1, with a binary response, we have $n_0 + n_1$ experimental units. At random n_0 are assigned to T_0 and the remainder to T_1; then a binary response is recorded for each individual. The randomization null hypothesis is that the response on any unit is the same whichever treatment is applied to that (or indeed other) units, a strong deterministic null hypothesis. Under that hypothesis we can reconstruct the observations that would have been obtained under any other treatment allocation and hence we can reconstruct also the values of R_0 and R_1, the numbers of successes in the two treatment groups that would have been obtained. In all these hypothetical tables the value of $R_0 + R_1$ is the same, equal to its observed value $r_.$. In fact in this distribution R_1 corresponds to the number of successes when a sample of size n_1 is drawn randomly without replacement from a finite population of n individuals containing r successes and hence has the hypergeometric distribution.

The non-null randomization distribution theory is difficult (Copas, 1973); so far as we know, the non-central distribution cannot be given a direct interpretation in this context.

4.5.3 A more general case

We now deal briefly with the general linear logistic model in which the independent binary random variables $Y_1, \ldots, Y_n$ with $E(Y_j) = \theta_j$ are such that

$$\lambda_j = \log \{\theta_j/(1 - \theta_j)\} = x_j\beta \qquad (4.31)$$

leading to the likelihood

$$\prod \frac{\exp(x_j \beta y_j)}{1 + \exp(x_j \beta)}. \tag{4.32}$$

To form a realistic Poisson-based model we would probably aggregate by merging any groups of individuals with the same vector of explanatory variables. Here, however, we shall not do this but rather shall treat each individual separately. For this, we replace each Y_j by two independent Poisson variables (Z_{j0}, Z_{j1}) with means (μ_{j0}, μ_{j1}); write $\mu_{j.} = \mu_{j0} + \mu_{j1}$, $Z_{j.} = Z_{j0} + Z_{j1}$. The corresponding likelihood can then be written as before:

$$\prod \frac{e^{-\mu_{j.}} \cdot \mu_{j.}^{z_{j.}}}{z_{j.}!} \binom{z_{j.}}{z_{j0}} \left(\frac{\mu_{j0}}{\mu_{j.}}\right)^{z_{j0}} \left(\frac{\mu_{j1}}{\mu_{j.}}\right)^{z_{j1}}. \tag{4.33}$$

Consider the log linear model

$$\log \mu_{j0} = \alpha_j, \qquad \log \mu_{j1} = \alpha_j + x_j \beta,$$
$$\mu_{j0}/\mu_{j.} = 1/(1 + e^{x_j \beta}), \qquad \mu_{j1}/\mu_{j.} = e^{x_j \beta}/(1 + e^{x_j \beta}). \tag{4.34}$$

Now take the special case $z_{j.} = 1$ with $y_j = 0$ corresponding to $z_{j0} = 1$, $z_{j1} = 0$ and $y_j = 1$ corresponding to $z_{j0} = 0$, $z_{j1} = 1$. Then the second factor in (4.33) is exactly the binary linear logistic likelihood (4.32) and the first factor depends only on the nuisance parameter $\mu_{j.}$. Therefore the conclusions of Section 4.5.1 extend immediately to this general case.

For the multinomial model, instead of conditioning as in (4.33) on $\{z_1, \ldots, z_n.\}$, we condition only on $z_{..} = \Sigma z_{j.}$, the total number of successes. Thus we replace the right-hand side of (4.33) by

$$\frac{e^{-\mu_{..}} \mu_{..}^{z_{..}}}{z_{..}!} \frac{z_{..}!}{\prod z_{j0}! \pi_{j1}!} \prod \phi_{j0}^{z_{j0}} \prod \phi_{j1}^{z_{j1}}, \tag{4.35}$$

where $\phi_{j0} = \mu_{j0}/\mu_{..}$, $\phi_{j1} = \mu_{j1}/\mu_{..}$, and the second factor in (4.35) specifies the multinomial distribution obtained from $z_{..}$ independent trials distributed among $2n$ cells. We now take the log linear model

$$\log \phi_{j0} = \varepsilon + \alpha_j, \qquad \log \phi_{j1} = \varepsilon + \alpha_j + x_j \beta_j, \tag{4.36}$$

where ε is a normalizing constant, so that

$$e^\varepsilon \sum_j e^{\alpha_j} (1 + e^{x_j \beta}) = 1.$$

It now follows in the usual notation that the second factor in (4.35) can itself be factorized as

$$\frac{z_{..}!}{\prod z_{j0}! \prod z_{j1}!} \frac{e^{\sum \alpha_j z_j.} \cdot e^{\sum x_j \beta z_{j1}}}{\sum e^{\alpha_j z_j.} \cdot (1 + e^{x_j \beta z_{j1}})}. \tag{4.37}$$

If we now rewrite (4.35) in terms of $\phi_1., \ldots, \phi_n., \beta$ we see that (4.37) corresponds to the distribution of the marginal totals, in fact fixed at 1 in our context, multiplied by the conditional distribution given the marginal totals, which is exactly the linear logistic likelihood (4.32). Note especially that in (4.37), as in (4.33), we have ensured that the combined likelihood is a product of factors, with β confined to one factor. So long as interest is concentrated on β only that factor need be considered.

As for the Poisson model we can also consider an aggregated model in which individuals with the same vector of explanatory variables are considered together.

While all these models are on occasion appropriate, we repeat that in studying the relation between a binary response and explanatory variables it is unnecessary and potentially confusing to model anything other than the conditional distribution of the responses for fixed values of the explanatory variables.

Example 4.3 Educational plans of Wisconsin schoolboys (continued)
A main effects logistic model was fitted in Example 2.18. The equivalent log-linear model, in addition to the main effects SES, IQ, and PE, contains the three-factor interaction SES × IQ × PE and the three corresponding two-factor interactions (to give the correct three-way marginal frequencies) plus the three interactions of the main effects with CP. Interpretation of the main effects is then given by differences across CP of the interaction parameters. For example, the estimated parameters for the SES × CP interaction in the log-linear model (using GLIM, for which the parametrization is particularly convenient in this instance as it sets the first parameter of each subset to zero), are

SES:		L	LM	UM	H
CP:	Yes	0	0	0	0
	No	0	−0.36	−0.66	−1.41

and differencing across CP gives the logistic model estimates for SES as in Table 2.16. The standard errors and fitted frequencies given by the two formulations also agree.

Bibliographic notes

A thorough account of case-control studies and epidemiological applications is in the book by Breslow and Day (1980); for an econometric viewpoint, see Amemiya (1985). Prentice and Pyke (1979) discuss theoretical equivalences.

The formal connection between discriminant analysis and logistic regression was probably known to early workers in these subjects; see, for example, Cornfield (1962). For a qualitative discussion of the distinctions, see Cox (1966b) and for quantitative evaluation of relative efficiency Efron (1975) and Ruiz (1989); for the relevance to testing goodness of fit, see Kay and Little (1987).

The interconnections of Poisson and binomial or multinomial distributions were exploited to simplify distributional calculations by Fisher (1922).

The extensive literature on log linear models, making no distinction between response and explanatory variables, can be approached via the books of Bishop, Fienberg and Holland (1975) and Fienberg (1977). To study a long series of influential papers by Goodman, start with his 1985 review.

CHAPTER 5

More complex responses

5.1 Introduction

The central theme of this book is the analysis of data in which there is a single response variable, binary in character. Yet it is clear that many of the topics and methods discussed in previous chapters can be generalized to other kinds of response variable. In the present chapter we outline some of the more immediate generalizations, namely to

1. paired preference data, in which each response is not 'success' or 'failure' but rather an indication of preference between two conditions or stimuli;
2. nominal data, where the response is one of a number of qualitatively different possibilities;
3. ordinal data, in which the qualitatively different possibilities are in some natural order;
4. multivariate binary data;
5. mixed binary and continuous responses.

We outline below some of the methods for the analysis of such data, concentrating on possibilities fairly closely linked with those for binary data.

5.2 Paired preferences

Suppose that there are m objects or treatments $A_1, \ldots, A_m$ for comparison, and that the observations consist of expressions of preference between pairs of objects; out of n_{st} comparisons of A_s with A_t, let A_s be preferred r_{st} times and A_t be preferred $r_{ts} = n_{st} - r_{st}$ times. The observations are thus binary, although concerned with responses of different types in different sections of the data.

Denote by θ_{st} the probability that A_s is preferred to A_t. We may hope to be able to express these $\frac{1}{2}m(m-1)$ parameters in terms of a much smaller number; there are several approaches to this. With a lot of data we can try to construct a set of m points $P_1, \ldots, P_m$ in a small number d of dimensions such that the relative positions of P_s and P_t determine θ_{st}. With limited data it may be more profitable to assume provisionally that a one-dimensional comparison is involved, i.e. that each object A_s is characterized by a parameter ρ_s and that θ_{st} is a simple function of $\rho_s - \rho_t$. Thus we write

$$\theta_{st} = g(\rho_s - \rho_t), \quad \theta_{ts} = 1 - \theta_{st} = g(\rho_t - \rho_s),$$

so that $g(x) = 1 - g(-x)$ and $g(0) = \frac{1}{2}$. Equivalently the logistic transform is

$$\lambda_{st} = \log\left(\frac{\theta_{st}}{\theta_{ts}}\right) = \log\left(\frac{g(\rho_s - \rho_t)}{1 - g(\rho_s - \rho_t)}\right) = f(\rho_s - \rho_t). \quad (5.1)$$

say, where $f(0) = 0$. The simplest form for $f(x)$ is a linear function, $f(x) = ax$, and without loss of generality we may take $a = 1$, this being equivalent to a choice of units on the axis on which the ρ's are measured. This leads us to consider

$$\lambda_{st} = \rho_s - \rho_t, \quad \theta_{st} = \frac{e^{\rho_s}}{e^{\rho_s} + e^{\rho_t}} = \frac{\gamma_s}{\gamma_s + \gamma_t}, \quad (5.2)$$

where $\gamma_s = \log \rho_s$.

The representation (5.2) is called the Bradley–Terry model; see the bibliographic notes for references on the development of the model.

The normal-theory analogue of (5.2) concerns an incomplete block design with two plots per block. The within-block analysis of this consists, in effect, of a normal-theory linear model for the differences of the two observations in a block, specifying an expectation $\rho_s - \rho_t$ for a difference between the sth and tth treatments.

The parameters ρ_s and ρ_t may themselves be linear combinations of other parameters, as in Example 5.1 below, where the m treatments for comparison are the 2^3 treatments of a factorial design. Likewise an effect corresponding to the order, within the pair, of presentation for testing may be represented by writing

$$\lambda_{st} = \rho_s - \rho_t \pm \delta, \quad (5.3)$$

where $+\delta$ is taken if A_s is tested before A_t, and $-\delta$ if A_t before A_s. Then we have

$$\theta_{st} = e^{\rho_s - \rho_t \pm \delta}/(1 + e^{\rho_s - \rho_t \pm \delta})$$

or, equivalently, θ_{st} is equal to

$$\kappa\gamma_s/(\kappa\gamma_s + \gamma_t) \quad \text{or} \quad \gamma_s/(\gamma_s + \kappa\gamma_t), \tag{5.4}$$

where $\delta = \log \kappa$, according to whether A_s is tested before A_t, or vice versa. In this form the parameter κ represents a multiplicative order effect as introduced by Davidson and Beaver (1977). Provided the model for λ_{st} remains linear in any unknown parameters it may be fitted by the standard methods for a linear logistic model.

The model (5.2) can fail in at least two ways. First, the objects may be arranged one-dimensionally but the function $f(\cdot)$ in (5.1) may be non-linear. This suggests the replacement of (5.2) by

$$\lambda_{st} = \rho_s - \rho_t + \varepsilon(\rho_s - \rho_t)^3, \tag{5.5}$$

followed by the estimation of ε, for example by maximum likelihood methods. The second possibility is that the comparison of the objects is essentially more complex than the comparison of objects arranged in one dimension.

The above discussion assumes a simple binary response – preference for A_s or A_t. When a response of no preference is also permitted we have three response categories: prefer A_s, no preference and prefer A_t. If an underlying latent response distribution is assumed, a logistic model may be formulated and fitted. This is discussed in Section 5.4, and Example 5.2 analyses some data with both a no-preference category and also an order effect.

Example 5.1 Preference test on coffee (Bradley and El-Helbawy, 1976)
Two brands of coffee were tested, each at two brew strengths and two roast colours. The treatments for comparison thus represent the eight combinations in a 2^3 factorial design. It seems likely that 26 judges each assessed 28 treatment combinations once, although this is not entirely clear. A more thorough analysis would take into account possible inter-judge differences and this we do not have the information to do.

The parameters ρ_s ($s = 1,\ldots,8$) can be written as a linear combination of the main effects and interaction terms of the 2^3 design, i.e.

$$\rho_s = \beta_A x_{A,s} + \beta_B x_{B,s} + \beta_C x_{C,s} + \beta_{AB} x_{A,s} x_{B,s} + \beta_{BC} x_{B,s} x_{C,s}$$
$$+ \beta_{AC} x_{A,s} x_{C,s} + \beta_{ABC} x_{A,s} x_{B,s} x_{C,s}, \tag{5.6}$$

where the x's take values ± 1 representing the two levels of factors A

Table 5.1 *Data for preference test on coffee (showing also fitted frequencies based on model containing A, B, C and AC)*

Pair	$x_{A,s}$ $x_{B,s}$ $x_{C,s}$	$x_{A,t}$ $x_{B,t}$ $x_{C,t}$	Preference for S out of 26 Observed	Fitted
1	-1 -1 1	-1 -1 -1	11	10.4
2	-1 1 -1	-1 -1 -1	11	11.6
3	-1 1 1	-1 -1 -1	10	9.1
4	1 -1 -1	-1 -1 -1	7	8.7
5	1 -1 1	-1 -1 -1	12	10.9
6	1 1 -1	-1 -1 -1	7	7.6
7	1 1 1	-1 -1 -1	10	9.6
8	-1 1 -1	-1 -1 1	16	14.3
9	-1 1 1	-1 -1 1	11	11.6
10	1 -1 -1	-1 -1 1	11	11.1
11	1 -1 1	-1 -1 1	12	13.5
12	1 1 -1	-1 -1 1	11	9.8
13	1 1 1	-1 -1 1	14	12.2
14	-1 1 1	-1 1 -1	11	10.4
15	1 -1 -1	-1 1 -1	11	9.9
16	1 -1 1	-1 1 -1	12	12.2
17	1 1 -1	-1 1 -1	8	8.7
18	1 1 1	-1 1 -1	11	10.9
19	1 -1 -1	-1 1 1	12	12.5
20	1 -1 1	-1 1 1	15	14.9
21	1 1 -1	-1 1 1	11	11.1
22	1 1 1	-1 1 1	13	13.5
23	1 -1 1	1 -1 -1	17	15.4
24	1 1 -1	1 -1 -1	12	11.6
25	1 1 1	1 -1 -1	13	14.0
26	1 1 -1	1 -1 1	10	9.4
27	1 1 1	1 -1 1	8	11.6
28	1 1 1	1 1 -1	14	15.4

A, brew strength; B, roast colour; C, brand.

Table 5.2 *Preference test on coffee: estimated parameters*

Factor	Estimate	Factor	Estimate
A	-0.154	AB	-0.019
B	-0.104	AC	0.193
C	-0.010	BC	-0.024
		ABC	-0.040

All estimates have standard error 0.050.

(brew strength), B (roast colour) and C (brand). The data are shown in Table 5.1.

The estimated parameters are given in Table 5.2. The main effects A and B are significant and so is the AC interaction. Refitting the model, omitting all nonsignificant interaction terms, does not change the values of the estimates to the order of accuracy quoted; this is because of near orthogonality. The likelihood ratio statistic for testing the fit of the factorial model (5.6) gives $\chi^2 = 5.17$, with 21 degrees of freedom; for the model omitting non-significant interactions, but containing A, B, C and AC, it is $\chi^2 = 6.18$ with 24 degrees of freedom. Note that the numbers of trials in the different cells are sufficient for a likelihood test based on a full fit to be justified. Such a good fit gives reason to suspect the independence assumptions underlying analysis of the data; for example, have systematic judge effects biased the residual error? Without further knowledge we can interpret the fitted model only cautiously as suggesting the following. The preferred level for brew strength is $x_A = -1$ and for roast colour is $x_B = -1$. To interpret the significant AC interaction we look at those pairs for which $x_{A,s}x_{C,s} - x_{A,t}x_{C,t} = \pm 2$. Of the four possibilities, one shows a marked preference: in the 104 comparisons for which both samples are of brand $x_C = -1$, the preferences for brew strength $x_A = -1$ to $x_A = 1$ are 71 to 33. The fitted frequencies based on a model containing the main effects A, B, C and the interaction AC agree closely with the observed frequencies (Table 5.1).

5.3 Nominal data

We have so far concentrated on data in which there are just two possible values corresponding to each response variable. We now consider rather briefly how to proceed if there are $h + 1$ possible values, labelled $0, \ldots, h$, these being regarded as unstructured, initially at least. For example we do not regard the responses as ordered or as formed from a cross-classified combination of more basic responses. Examples are eye or hair colour as a response, choice between different educational establishments or courses, choice between different methods of transport to work, and so on.

Typically, in addition to the response, there are explanatory variables for each individual.

If h is large, and especially if it is so large that some of the cells have very low frequencies in the data under analysis, there are major difficulties and an initial step is often to try and amalgamate categories of response in some natural way. We concentrate here on situations where h is reasonably small. There are even here a number of different routes to interpretation. Unless there is clear prior reason for preferring one approach it may be necessary to do several analyses and to choose between the interpretations on subject-matter grounds. Here are some possibilities.

1. *Conditional nested variables.* It may be sensible to represent the variables by a nested structure of binary variables defined conditionally. As a simple example, suppose that there are three possible responses: survived, dead from cause A, dead from cause B. We might then define Y_1 as 0 for survivors and 1 for those that die, and then conditionally on death, i.e. on $Y_1 = 1$, define $Y_{1,2}$ as 0 for death from cause A and as 1 for death from cause B. More generally, there would be an initial binary variable Y_1, and conditionally on $Y_1 = 0$ a binary variable $Y_{0,2}$ and conditionally on $Y_1 = 1$ a binary variable $Y_{1,2}$ with possibly four more binary variables $Y_{00,3}$, $Y_{01,3}$, $Y_{10,3}$ and $Y_{11,3}$, and so on, with some of the variables possibly not being necessary, as in our example where cause of death is not defined for survivors! Justifications for such subdivisions could be either that a few of the variables could be regarded as more important than others on subject-matter grounds or that dependence on explanatory variables could be captured in one or two of the components. In the latter case it might be possible, at least in principle, to determine a suitable nested structure for defining the variables purely empirically, at least as a pointer to more detailed analysis. On the whole, however, if this approach is to be adopted it seems wise to introduce specific subject-matter considerations at the start.

2. *Cross-classified component variables.* This is rather similar in spirit to (1) except that the derived variables are cross-classified rather than conditionally nested. Thus if $h + 1 = 2^m$, we might attempt to define m binary variables so that the responses are identified via the multivariate structure of the m binary variables. Now formally this can always be done, indeed in many different ways, and the objective is to define the component variables preferably in a way that both has some subject-matter interpretation and produces simple empirical description of the dependence on explanatory variables. Thus it might be possible to arrange that the dependence on explanatory variables is confined to a small number of components. Constructive empirical

ways of finding such variables could be based on careful inspection of the results of fitting the generalized logistic model described below, although perhaps more commonly subject-matter considerations would be a better guide.

3. *Assignment of scores.* A quite different approach is to aim at scores for the $h + 1$ categories of response chosen to represent the dependence on explanatory variables as strongly as possible. A fairly simple way is to assign arbitrary scores and then to maximize the squared multiple correlation coefficient for the multiple linear regression of these scores on the explanatory variables. As with other applications of what is essentially canonical regression analysis, it may be possible to 'explain' a quite high proportion of the dependence present, but firm interpretation of the 'scores', i.e. of the form of the coefficients of the canonical variable, may not be possible.

4. *Generalized logistic models.* In the light of our previous emphasis on logistic models it is natural to generalize the linear logistic model to problems with more than two possible responses. To do this we can take a linear logistic relation for the comparison of the probabilities of any two levels of response. Thus if the possible responses are labelled arbitrarily $0, 1, \ldots, h$, regression on a variable x can be represented by

$$\log \left\{ \frac{\mathrm{prob}\,(Y = a)}{\mathrm{prob}\,(Y = b)} \right\} = (\alpha_a - \alpha_b) + x(\beta_a - \beta_b). \tag{5.7}$$

This is equivalent to

$$\mathrm{prob}\,(Y = a) = \frac{e^{\alpha_a + x\beta_a}}{\sum_{b=0}^{h} e^{\alpha_b + x\beta_b}} \qquad (a = 0, 1, \ldots, h). \tag{5.8}$$

Two constraints can be imposed to make the parameters unique; for example, we may take

$$\alpha_0 = \beta_0 = 0. \tag{5.9}$$

There is an obvious generalization when there are several explanatory variables.

The model does, however, have the special property of what is sometimes called independence from irrelevant alternatives. Take the above model for $h + 1$ possible responses and suppose one of the responses, say $Y = h$, is split at random with fixed probability into either $Y = h$ or $Y = h + 1$, i.e. the number of possible response categories is increased in an irrelevant fashion. It is easily seen that under the present model the regression coefficients β are unchanged. This point has obvious relevance to the combining of response categories, the distinction between which appears to be irrelevant. As an example, suppose that the response categories are methods of travel to work and that $Y = h$ corresponds to travel by bicycle. Suppose then that a distinction is introduced between black bicycles ($Y = h$) and non-black bicycles ($Y = h + 1$). For many purposes one might expect this to be an irrelevant split, although of course the possibility could not be totally dismissed that colour is a surrogate for some meaningful aspect of the individuals under study. One might expect, however, that the regression coefficients on explanatory variables should not be affected by such a split.

A preliminary analysis and test of the model can be based on (5.7). With grouped data a series of linear plots on a logistic scale should be obtained. If one response is relatively frequent it will often be convenient to call it the zero response, to apply (5.9) and to take $\alpha = 0$ in all comparisons (5.7). We thus obtain h lines on a logistic scale.

The model can be fitted by maximum likelihood in a straightforward way. If it is required to use the empirical logistic transform, a generalization is needed. Consider grouped data and suppose that in one group there are m trials, that the probabilities of the $h + 1$ responses are $\theta_{(0)}, \ldots, \theta_{(h)}$ and that the corresponding numbers of responses are $R_{(0)}, \ldots, R_{(h)}$, which are multinomially distributed. Asymptotically

$$\left.\begin{aligned}
E(\log R_{(a)}) &= \log m + \log \theta_{(a)}, \\
\mathrm{var}\,(\log R_{(a)}) &= \frac{1 - \theta_{(a)}}{m\theta_{(a)}} \sim \frac{1}{R_{(a)}} - \frac{1}{m}, \\
\mathrm{cov}\,(\log R_{(a)}, \log R_{(b)}) &\sim -\frac{1}{m} \qquad (a \neq b).
\end{aligned}\right\} \tag{5.10}$$

Of the $(h + 1)$ frequencies $R_{(a)}$, only h are independent and there is no loss of information in replacing the $\log R_{(a)}$'s by h contrasts which may, for example, be

$$Z_{(a0)} = \log R_{(a)} - \log R_{(0)} \qquad (a = 1, \ldots, h).$$

These have asymptotic expectations

$$\log (\theta_{(a)}/\theta_{(0)}) \tag{5.11}$$

and also

$$\text{var}(Z_{(a0)}) = \text{var}(\log R_{(a)}) - 2\,\text{cov}(\log R_{(a)}, \log R_{(0)})$$
$$+ \text{var}\{\log R_{(0)}\} \sim R_{(a)}^{-1} + R_{(0)}^{-1}, \qquad (5.12)$$
$$\text{cov}(Z_{(a0)}, Z_{(b0)}) \sim R_{(0)}^{-1} \qquad (a \neq b).$$

Now the generalized models (5.7) and (5.8) specify a linear relation for the asymptotic expectations (5.11) and, the covariance matrix being estimated from (5.12), the method of generalized least squares can be applied for estimation and testing. This approach is set out in detail for the analysis of contingency tables by Plackett (1962) and Goodman (1963). If $h = 1$, so that the response is binary, the methods reduce to the use of the empirical logistic transform.

More explicitly, a vector of h components $Z_{(0)}$ can be formed from each grouped observation and with k groups of data, a vector Z of hk components is thereby defined. Its asymptotic expectation is, by (5.7) and (5.11), linear in the unknown parameters, say

$$E(Z) \sim \alpha\beta. \qquad (5.13)$$

Further the covariance matrix of z is estimated as

$$V = \text{diag}(V_1, \ldots, V_k), \qquad (5.14)$$

where V_j is the $h \times h$ matrix obtained by applying (5.12) to the jth group of observations. Thus β is estimated by

$$\tilde{\beta} = (a^{\mathrm{T}}V^{-1}a)^{-1}a^{\mathrm{T}}V^{-1}Z; \qquad (5.15)$$

the covariance matrix of $\tilde{\beta}$ is

$$(a^{\mathrm{T}}V^{-1}a)^{-1}. \qquad (5.16)$$

The residual sum of squares is

$$(Z - a\tilde{\beta})^{\mathrm{T}}V^{-1}(Z - a\tilde{\beta}) \qquad (5.17)$$

and gives a test of the adequacy of the model using the chi-squared distribution with $(hk - p)$ degrees of freedom, where p is the number of parameters fitted in (5.13).

If in some groups the frequencies of certain responses are low, it will be necessary to omit the corresponding components from Z, or to use instead the method of maximum likelihood.

5.4 Ordinal data

In some situations response may be measured on an ordinal scale. For instance, in a bioassay animals may be classified after test as unaffected, seriously affected or dead; in testing a food product each judge may classify the product as very good, good, fair, poor or very poor. We label the responses conventionally by $Y = 0, 1, \ldots, h$ but without any implication that the differences between adjacent categories are in any meaningful sense the same. Note that in designing studies involving ordinal responses it is often important to 'anchor' the scale division points as firmly as feasible by guidance on the interpretation to be given to the verbal descriptions.

There are a number of possible approaches to the analysis of ordinal data.

For example, we might initially treat the $h + 1$ response categories as nominal using methods which ignore their order and examine the relation between the estimates and the ordering of the categories. This was effectively the approach in Example 2.18, although there the factor SES with four ordered categories was an explanatory factor, not a response variable.

Alternatively, simple scores based on some assumed notional underlying latent scale can be tried. If this is done it is important to check in some way the sensitivity of any conclusions to changes in the values of the scores assigned.

A more formal approach to the analysis of ordinal data can be motivated via the idea of a continuously distributed latent response variable U from which the ordinal response Y is produced via 'cut-off' points $\xi_0, \ldots, \xi_{h-1}$:

if and only if $U \leqslant \xi_0$, then $Y = 0$;
if and only if $\xi_0 < U \leqslant \xi_1$, then $Y = 1$;
$$\vdots$$
if and only if $\xi_{h-1} < U$, then $Y = h$.

Thus,

$$\text{prob}(Y \leqslant a) = \text{prob}(U \leqslant \xi_a) \qquad (a = 0, \ldots, h - 1).$$

Models are now produced in effect in two stages:

1. by the convention that under some standard baseline conditions, U has some convenient distribution, e.g. the standard logistic or the standard normal;

2. by the assumptions that the effect of explanatory variables is represented by a simple change in the distribution of U, usually a shift in location.

If under baseline conditions U has the standard logistic distribution, then

$$\theta_a = \text{prob}\,(Y \leqslant a) = \text{prob}\,(U \leqslant \xi_a) = e^{\xi_a}/(1 + e^{\xi_a}),$$
$$\log\,[\theta_a/(1 - \theta_a)] = \xi_a.$$

If for the individual with row vector of explanatory variables x_i the shift in the mean of the logistic distribution is $x_i\beta$, then for the ith individual with response Y_i

$$\theta_{ia} = \text{prob}\,(Y_i \leqslant a) = e^{\xi_a + x_i\beta}/(1 + e^{\xi_a + x_i\beta}),$$

so that

$$\log\,[\theta_{ia}/(1 - \theta_{ia})] = \xi_a + x_i\beta. \tag{5.18}$$

Interest lies in the estimation of β regarding $\xi_0,\ldots,\xi_{h-1}$ as nuisance parameters.

In particular, for a $(h + 1) \times 2$ contingency table,

$$\log\,[\theta_{0a}/(1 - \theta_{0a})] = \xi_a, \quad \log\,[\theta_{1a}/(1 - \theta_{1a})] = \xi_a + \beta,$$

i.e.

$$\theta_{1a}/(1 - \theta_{1a}) = e^{\beta}\theta_{0a}/(1 - \theta_{0a}) \qquad (a = 0,\ldots,h - 1).$$

Here e^{β} represents the odds ratio and (5.18) is known as the proportional odds model. Note that reversal of the ordinal scale involves only a change of sign of β; also the same form of model holds if adjacent categories are combined. A disadvantage of (5.18), however, is the computational effort required if many parameters are involved.

If U is assumed to follow an extreme value distribution then $\log\,[\theta_{ia}/(1 - \theta_{ia})]$ in (5.18) is replaced by $\log(-\log\theta_{ia})$ or $\log\,[-\log(1 - \theta_{ia})]$, depending whether a minimum or maximum response is implied.

Example 5.2 Preference testing experiment on food mixes
(Davidson and Beaver, 1977)
Four packaged food mixes were compared, all possible pairs being tested. Each pair was presented for testing in the order (s, t) and

(t, s). Respondents were asked to reply on the 3-point scale: prefer s, no preference, prefer t. The data are given in Table 5.3.

We do not know whether the same judges compared each pair and so cannot take account of any inter-judge differences in our analysis.

We assume, as in Section 5.2, that each food mix is characterized by a parameter ρ_s ($s = 1, \ldots, 4$) and the preference between any pair (s, t) depends upon $\rho_s - \rho_t$. From (5.3) and (5.18), we have

$$\log \left[\theta_{st,a} / (1 - \theta_{st,a}) \right] = \xi_a \pm \delta + \rho_s - \rho_t \qquad (a = 0, 1), \quad (5.19)$$

where $+ \delta$ is taken if the order of testing is (s, t) and $- \delta$ if (t, s).

Table 5.3 *Frequency of response for preference testing of food mixes*

Pair (s, t)	Order	Frequency of response			
		Prefer s	No preference	Prefer t	Total
1, 2	1, 2	23	8	11	42
	2, 1	6	8	29	43
1, 3	1, 3	27	5	11	43
	3, 1	14	6	22	42
1, 4	1, 4	35	1	6	42
	4, 1	11	4	27	42
2, 3	2, 3	34	1	6	41
	3, 2	16	3	23	42
2, 4	2, 4	29	2	9	40
	4, 2	15	5	22	42
3, 4	3, 4	26	5	11	42
	4, 3	14	5	24	43

Table 5.4 *Maximum likelihood estimates and asymptotic standard errors for preference testing of food mixes*

Parameter	Estimate	Standard error
ξ_0	-0.195	0.227
ξ_1	0.314	0.227
δ	0.840	0.094
ρ_1	0.240	0.373
ρ_2	0.525	0.276
ρ_3	0.037	0.196
ρ_4	0	—

The GLIM program sets $\rho_4 = 0$.

Maximum likelihood estimates of the parameters in (5.19) are given in Table 5.4; these were obtained using the GLIM program given by Hutchison (1985). The order effect is highly significant ($\hat{\delta} = 0.840$, with standard error 0.094), with preference in favour of the first mix tested, as indeed is obvious from the data. Food mix 2 ($\hat{\rho}_2 = 0.525$, with standard error 0.276) is the most preferred of the four mixes. The data are fitted adequately by the model (5.19); the likelihood ratio test gives $\chi^2 = 19.9$ with 18 degrees of freedom, this test being reasonably justified in this case.

5.5 Multivariate binary data

We can apply the model of Section 5.3 when the responses have a factorial structure. The simplest situation of this kind occurs when there are four levels of response corresponding to a pair (Y_1, Y_2) of binary variables. The four responses can thus be labelled $(0, 0)$; $(0, 1)$; $(1, 0)$; $(1, 1)$. In (5.8) we can write the probabilities as proportional to

$$1; \qquad e^{\alpha_{01} + x\beta_{01}}; \qquad e^{\alpha_{10} + x\beta_{10}}; \qquad e^{\alpha_{11} + x\beta_{11}}. \qquad (5.20)$$

There are corresponding formulae when the factorial structure is more complicated.

One property of (5.20) is that the marginal dependence of Y_1 on x is logistic if and only if $\alpha_{01} + \alpha_{10} = \alpha_{11}$ and $\beta_{01} + \beta_{10} = \beta_{11}$; and this is the condition for the independence of Y_1 and Y_2. When the second condition is satisfied but not the first, there are parallel conditional logistic relations between, say, Y_1 and x given (i) $Y_2 = 0$ and (ii) $Y_2 = 1$.

If (5.20), or a generalization of it, is fitted to data, it is often helpful to try to describe the α's and the β's in terms of fewer than the full number of parameters. This is broadly analogous to the condensation of the results of factorial experiments in terms of main effects and low-order interactions. In fact one way in which condensation can be attempted in the present case is to replace the α's and β's by a smaller number of contrasts. Another possibility is that sets of, say, β's are equal, i.e. that there are groups of responses for which (5.7) is independent of x. In general if there are q binary response variables and p parameters corresponding to regressor variables, we shall in effect obtain in the model p sets of 2^q parameters, only contrasts within any particular set being meaningful in view of (5.9). Although the parameters will be estimated with different precisions, it is legitimate to apply the semi-normal plotting technique (Daniel, 1959) to each of the p sets of estimates as a guide to their condensation.

The objective is to describe the regression of the Y's on the x's as concisely as possible and this can be compared with the aim of canonical correlation and canonical regression analysis in the multivariate analysis of quantitative responses. There the search is for linear combinations of the response variables which have high correlation with the explanatory variables. The same idea could formally be applied here, although it is not clear in general that linear combinations of correlated binary variables have particular virtue.

Indeed a natural way of deriving new binary variables from a given set involves not linear transformations, but permutations, in a way best illustrated via a simple example. Suppose that we have two binary variables

$$Y_1 = \begin{cases} 1, \text{ wife votes labour,} \\ 0, \text{ wife does not vote labour,} \end{cases}$$

$$Y_2 = \begin{cases} 1, \text{ husband votes labour,} \\ 0, \text{ husband does not vote labour.} \end{cases}$$

The four possible outcomes can be described also via

$$Y_1' = \begin{cases} 1, \text{ wife and husband agree } (Y_1 = Y_2), \\ 0, \text{ wife and husband disagree } (Y_1 \neq Y_2), \end{cases}$$

$$Y_2' = Y_1$$

or via

$$Y_1'' = Y_1', \qquad Y_2'' = Y_2.$$

That is, there are three distinct 'coordinate' systems for specifying the four possible outcomes. It is possible that for describing the joint distribution and its dependence on explanatory variables one of these systems leads to more easily interpreted conclusions: for example, if Y_1'' and Y_2'' are independent, the other representations will not have independent components.

Bloomfield (1974) suggested that the transformations of coordinates should be restricted to those linear in arithmetic mod 2. Note, for instance, that in the above examples $Y_1' = Y_1 + Y_2$ mod 2. For $p > 3$ not all transformations of binary coordinates are linear.

A special case of (5.20) is when the x variable or variables take only two values. In particular with two groups and with two binary responses the probabilities are proportional to

Group 0   1, $e^{\alpha_{01}}$, $e^{\alpha_{10}}$, $e^{\alpha_{11}}$;

Group 1   1, $e^{\alpha_{01} + \Delta_{01}}$, $e^{\alpha_{10} + \Delta_{10}}$, $e^{\alpha_{11} + \Delta_{11}}$.

There is an extensive literature on significance tests in contingency tables in which some dimensions correspond to responses and some to explanatory variables; see, for example, the review of Lewis (1962). Such tests are a valuable preliminary to the concise description of data. Another special case is when the linear model represents a linear, or possibly quadratic, surface in some quantitative factor variables. We then have models of response surfaces with multivariate binary responses.

Throughout this discussion, the logistic form is central. Note, however, that a similar family of models can be formed with other functions than exponentials in (5.20) and that given enough data one could choose an appropriate function empirically.

A number of the problems of normal-theory multivariate analysis have fairly direct formal analogues for binary variables, but some caution is needed. Thus in the perhaps not very practical problem of testing the equality of the vector means of two normal samples, Hotelling's T^2, essentially equivalent to a discriminant function test, is the usual procedure and is certainly reasonable if an interpretation invariant under linear transformation of the components is required. But such invariance has, as noted above, little relevance for binary variables. Thus, while a direct analogue of Hotelling's T^2 can be formed, essentially by treating 0s and 1s as quantitative, it is often more sensible to examine the equality of the components one at a time and then, if an overall test is required, make a conservative allowance for selection, multiplying the smallest significance level by the number of components.

Finally, if there is a natural temporal or suspected causal ordering of the components, an analysis looking for conditional independencies will be sensible. We shall not discuss this large subject.

5.6 Mixed binary and continuous responses

When there are several response variables for each individual some may be binary and some continuous. For instance in a clinical trial one might measure depression in blood pressure and presence or absence of various symptoms. Now the use of several response measures does not on its own force the use of specifically multivariate techniques; it may be entirely adequate to do separate analyses one response at a time.

The situations where this is not the case and where some synthesis of the analyses for the joint variables is necessary are broadly as follows:

1. there may be many responses of roughly similar type and for ease of interpretation some reduction of dimensionality is needed, usually by the formation of new composite variables;
2. most of the variables may show little or no regression or treatment effect and one wants some assurance that the effects that are discovered are not accidents of selection, i.e. one wants a test of an overall null hypothesis of the absence of structure;
3. one may wish to interrelate treatment or regression effects on different response variables, perhaps showing that effects on some variables are in some sense largely explained by effects on other variables.

On the whole (1) is unlikely to arise with mixed continuous–binary response variables because these will usually be of rather different types; of course the combination of continuous and binary explanatory variables raises no special issues.

The second issue, that of allowance for selection, is normally adequately handled by the rule for independent responses; if p is the smallest significance level out of m variables tested, the significance level is to be taken as $1 - (1 - p)^m \simeq mp$. If this is too crude, because for example the variables are quite strongly dependent, more refined procedures can be based on some of the representations that follow.

The third and in many ways much the most interesting possibility is the exploration of the interrelation of the different responses and the implications for the interpretation of treatment and regression effects and the following discussion relates to that issue.

When we examine the joint distribution of binary and continuous responses, there are a number of routes to interpretation and the choice between them depends partly on subject-matter considerations and partly on which route leads to the simplest interpretation. Thus we may

1. represent the joint distribution in a way that treats binary and continuous components essentially symmetrically;
2. concentrate on the marginal distribution of the binary compo-

nents and the conditional distribution of the continuous components given the binary components, taking the continuous components collectively or one at a time;
3. concentrate on the marginal distribution of the continuous components and the conditional distribution of the binary components given the continuous components, taking the binary components collectively or one at a time.

In each case dependence on or independence from explanatory variables is of interest. Approach (2) is natural if the continuous variables are sensibly regarded as an intermediate response and (3) if the binary variables are so regarded.

For example, depression of blood pressure might in some contexts be regarded as an intermediate response and presence or absence of some overt symptoms as potentially consequent on blood-pressure change. In particular it might be relevant to consider whether any dependence of the latter on treatment and other explanatory variables can be explained via an effect on blood pressure. In another context some procedure might be classified as unsuccessful or successful and the distribution of the continuous variables studied separately within the two groups so defined.

Simple models for these three possible approaches are fairly easily formed from the components studied previously in the book. Such models will often be a convenient starting point for more detailed interpretation.

Thus a model that treats the variables essentially symmetrically can be formed from a multivariate normal distribution with linear regression on the explanatory variables, using the normal distribution to define a probit model for the discrete components via the notion of latent variables, as in Section 2.10. In this model, however, the conditional distribution of the continuous components given the binary components is not of particularly simple form, so that in particular it is not well suited to examining possible conditional independence of the continuous variables and the explanatory variables given the binary variables.

In the second type of formulation we would represent the dependence of binary variables on explanatory variables by one of the forms discussed earlier in the book. Then, given the binary variables, we can analyse the continuous variables by any appropriate standard method, in simple cases by normal-theory least squares. If the multivariate character of the continuous responses is important a first step is often to examine how the mean and covariance matrix depend on binary and explanatory variables.

In the third formulation the roles are reversed, the distribution of the continuous variables being studied first and then the conditional distribution of the binary responses. There are numerous possible complications that could arise in applications but the broad approach should be clear.

Bibliographic notes

The development of the Bradley–Terry model for paired preferences is reviewed by Bradley (1976) and an extensive bibliography is given by Davidson and Farquhar (1976). Many of the statistical problems associated with the model have been worked out in detail; an excellent account, dealing also with other methods for handling paired comparison data, is given by David (1988).

The proportional odds and proportional hazards models are applied to ordinal data by McCullagh (1980); see also McCullagh and Nelder (1983) and Agresti (1984). Methods based on individual logistic regressions are discussed by Anderson (1984) and by Begg and Gray (1984). Agresti (1988) uses loglinear models. A general review of methods applicable to ordinal and nominal data is given by Engel (1988).

For a brief discussion of multivariate binary data see Cox (1972); much of the related literature on multidimensional contingency tables is relevant. For an example see Montgomery, Richards and Braun (1986).

For a different approach to the analysis of ordinal data see Anderson and Philips (1981) and for a generalization of probit analysis Aitchison and Silvey (1957).

Correspondence analysis is described by Greenacre (1984) and Lebart, Morineau and Warwick (1984) and compared with other methods by Goodman (1986).

Methods for mixtures of continuous and discrete responses are studied in depth by Lauritzen and Wermuth (1989) and for a different approach, see Olkin and Tate (1961).

Theoretical background

A1.1 Introduction

It would hardly be sensible to aim in a short Appendix to cover in detail all the theoretical considerations underlying the methods set out in the book. Nevertheless such a key role is played by the methods of maximum likelihood and least squares, and by the relation between them, that an outline discussion of such matters seems a good idea. Our objective is to give the key ideas, mostly without proof.

Although it is not necessary to follow this route, we begin with an outline of the normal-theory linear model to be regarded as a special case of maximum likelihood theory.

A1.2 Ordinary least squares theory

Suppose that the $n \times 1$ vector of responses $y = (y_1, \ldots, y_n)^T$, not necessarily binary, is represented by a random vector $Y = (Y_1, \ldots, Y_n)^T$ such that

$$E(Y_k) = \sum_{s=0}^{p} x_{ks}\beta_s, \qquad E(Y) = x\beta, \qquad (A1.1)$$

where the $n \times d$ matrix x is known and β is a $d \times 1$ vector of unknown parameters, with $d = p + 1$. Equivalently

$$Y_k = \sum x_{ks}\beta_s + \varepsilon_k, \qquad Y = x\beta + \varepsilon, \qquad (A1.2)$$

where $E(\varepsilon) = 0$, and $\varepsilon = (\varepsilon_1, \ldots, \varepsilon_n)^T$ is an $n \times 1$ vector of errors.

Example A1.1
If $d = 1$, $x_{k0} = 1$, then $E(Y_k) = \beta_0$, $E(Y) = \beta_0 1$, representing a set of unstructured data.

Example A1.2
If $d = 2$, $x_{k0} = 1$, $x_{k1} = x_k$, then $E(Y_k) = \beta_0 + \beta_1 x_k$, representing linear

regression. If the term in β_0 is omitted we have a straight line through the origin.

Example A1.3
If $x_{x0} = 1$ and $x_{k1}, \ldots, x_{kp}$ represent values of p explanatory variables for the kth individual, we have multiple linear regression.

Example A1.4
Numerous extensions of Example A1.3 are possible in which derived explanatory variables, e.g. squares and products of the original variables, are added.

The least squares estimates are chosen to minimize

$$\sum_k \left(Y_k - \sum_s x_{ks}\beta_s \right)^2 = (Y - x\beta)(Y - x\beta)^{\mathrm{T}}$$

and therefore satisfy

$$\sum_s \left(\sum_k x_{kr}x_{ks} \right)\hat{\beta}_s = \sum_k x_{kr}Y_k, \qquad x^{\mathrm{T}}x\hat{\beta} = x^{\mathrm{T}}Y. \tag{A1.3}$$

Note that for comparison with maximum likelihood results it is convenient to record the partial derivative of the sum of squares with respect to β_r, times $(2\sigma^2)^{-1}$, this being

$$\sigma^{-2}\sum x_{kr}\left(Y_k - \sum x_{ks}\beta_s \right),$$

or in vector form

$$\sigma^{-2}(x^{\mathrm{T}}Y - x^{\mathrm{T}}x\beta).$$

The estimates are unbiased, $E(\hat{\beta}) = \beta$. Geometrically (A1.3) is best written in the form

$$x^{\mathrm{T}}(Y - x\hat{\beta}) = 0, \tag{A1.4}$$

corresponding to projecting Y orthogonally on to the space spanned by the columns of x; in the absence of error Y would lie exactly in that subspace and the true value β would be recovered.

The fitted values are obtained by replacing β in the defining model by their estimates $\hat{\beta}$, i.e. by

$$\hat{Y}_k = \sum x_{ks}\hat{\beta}_s, \qquad \hat{Y} = x\hat{\beta}.$$

The residuals are defined as differences of observed and fitted values, that is by

$$R_k = Y_k - \hat{Y}_k, \qquad R = Y - \hat{Y}. \qquad (A1.5)$$

The residual R_k should not be confused with the error ε_k, although often the two will be nearly equal.

To obtain further properties, either as a basis for confidence intervals or to justify this particular choice of estimates, further specification of the distribution of errors is needed. The two simplest sets of starting assumptions are as follows.

1. *Normal-theory assumptions.* The errors $\varepsilon_1, \ldots, \varepsilon_n$ are independently normally distributed with zero mean and variance σ^2.
2. *Second-order (quasi-likelihood) assumptions.* The errors have zero mean, are pairwise uncorrelated and have the same variance, σ^2.

Clearly the first set of assumptions imply the second and under both

$$E(\varepsilon_k \varepsilon_l) = 0 \quad (k \neq l), \qquad E(\varepsilon_k^2) = \sigma^2, \qquad E(\varepsilon \varepsilon^T) = \sigma^2 I. \qquad (A1.6)$$

Under (A1.6) the covariance matrix of the least squares estimates is

$$\text{cov}(\hat{\beta}) = (x^T x)^{-1} \sigma^2, \qquad (A1.7)$$

so that in particular the variance of the component $\hat{\beta}_s$ is σ^2 times the sth diagonal element of $(x^T x)^{-1}$, where we assume x is of full rank $d \leqslant n$.

Example A1.1 (continued)
When the model contains just the single parameter for the mean, $x^T x = n$, $(x^T x)^{-1} = 1/n$ and, the least squares estimate being the sample mean, the usual formula for the standard error of a mean is recovered.

To estimate σ^2 the sum of squares of the residuals is used, $S_{\text{res}} = \sum R_k^2 = R^T R$. Under the above assumptions

$$s_{\text{res}}^2 = S_{\text{res}}/(n - d) \qquad (A1.8)$$

is an unbiased estimate of σ^2.

A fairly common problem associated with these models arises when the parameter β is partitioned into two (or more) parts, $\beta^T = (\beta_1^T, \beta_2^T)$, where the dimensions of the components are d_1 and d_2, with $d = d_1 + d_2$. Suppose that it is required to test the agreement of the data with the simpler model with $\beta_2 = 0$. Denote by $S_{\text{res}}(\beta_1, \beta_2)$ and

$S_{res}(\beta_1, \beta_2 = 0)$ the residual sums of squares from the full model and from the restricted model. (Note that these are functions of the data not of the pseudo-arguments β_1, β_2.) The difference represents the improvement of fit achieved by allowing non-zero β_2. A formal procedure is based on the statistic

$$\frac{\{S_{res}(\beta_1, \beta_2 = 0) - S_{res}(\beta_1, \beta_2)\}/d_2}{S_{res}(\beta_1, \beta_2)/(n - d)}. \tag{A1.9}$$

Under normal-theory assumptions 'exact' distributions are available for the various statistics, in particular $\hat{\beta}$ having a multivariate normal distribution. Also strong theoretical optimal properties are available under the normal-theory assumptions, based essentially on the sufficiency of the statistics involved, whereas under second-order assumptions the resulting statistics are optimal only within very restricted families of procedures and the corresponding distributional results hold only as approximations.

A1.3 Extensions of least squares

To some extent the methods outlined in Section A1.2 remain reasonable even if the assumptions about the covariance of the errors are violated. If, however, the form of the covariance matrix is known, simple modifications of the least squares analysis are possible and these we now sketch.

A.1.3.1 Weighted least squares

Suppose that the errors are uncorrelated but of unequal variance with $\mathrm{var}(\varepsilon_k) = v_k\sigma^2$, where $v_1, \ldots, v_k$ are known constants. The simplest approach is to apply the previous discussion to the new response variables $Y_k/\sqrt{v_k}$ which have equal variance. Equivalently we minimize

$$\sum_k v_k^{-1}\left(Y_k - \sum_s x_{ks}\beta_s\right)^2.$$

The corresponding estimates satisfy, instead of (A1.3),

$$(x^T V^{-1} x)\hat{\beta}_v = x^T V^{-1} Y, \quad \mathrm{cov}(\hat{\beta}_v) = (x^T V^{-1} x)^{-1}\sigma^2, \tag{A1.10}$$

where $V = \mathrm{diag}(v_1, \ldots, v_n)$. Other properties are similarly modified. This broad method is called weighted least squares.

A1.3.2 Generalized least squares

Next we suppose that $\text{cov}(\varepsilon) = \sigma^2 V$, where V is a known matrix, not necessarily diagonal. Then the results (A1.10) continue to apply and can be regarded as derived via the method of ordinary least squares applied after a preliminary linear transformation to uncorrelated random variables of constant variance. This approach is called the method of generalized least squares. Geometrically, orthogonal projection in the Euclidean sense, as used in ordinary least squares, has been replaced by projection in the metric defined by the covariance matrix. This is seen most directly by writing the estimating equations in the form

$$x^T V^{-1}(Y - x\hat{\beta}_v) = 0.$$

A1.3.3 Weighted least squares with empirically estimated weights

Suppose now that the errors are uncorrelated, that $\text{var}(\varepsilon_k) = v_k \sigma^2$, but that the v_k are unknown and are estimated by consistent estimates $\tilde{v}_k$. Then in (A1.10) we can replace V by $\tilde{V} = \text{diag}(\tilde{v}_1, \ldots, \tilde{v}_n)$ to give weighted least squares estimates with empirically estimated weights.

A1.3.4 Weighted least squares with approximately linearized responses and empirically estimated weights

Sometimes instead of the Y_k being directly observed responses they are derived responses calculated in such a way that the linear representation for $E(Y_k)$ holds only as an approximation, even under the postulated model. We can still apply the method of weighted least squares and typically empirical estimation of weights will be needed.

Of course the last two methods involve approximations over and above the inevitable approximation involved in using a simple mathematical model to represent some phenomenon in the real world. A typical example of approximately linearized response variables are empirical logit transforms.

A1.4 Likelihood functions

As a basis for a more general treatment of parametric statistical inference, define the log likelihood function for an $n \times 1$ vector y of observed responses, represented by a vector random variable Y, by

$$l = l(\theta) = l(\theta, y) = \log f_Y(y; \theta), \qquad (A1.11)$$

where $f_Y(y; \theta)$ is the joint probability or probability as specified under the model, θ being an unknown $d \times 1$ vector parameter. To complete the specification we have to state the parameter space Ω_θ, the set of physically allowable values for θ in the application under study. For the random variable corresponding to (A1.11) we write $l(\theta; Y) = \log f_Y(Y; \theta)$.

In the log likelihood we may ignore any term that does not involve the unknown parameter.

The score function is defined for a scalar parameter θ by

$$u = u(y, \theta) = \partial l(\theta; y)/\partial \theta, \qquad (A1.12)$$

with a corresponding definition $U = u(Y, \theta)$ for a random variable. For a vector parameter θ the score is the $d \times 1$ vector of partial derivatives u defined by

$$u^T = (\partial/\partial\theta_1, \ldots, \partial/\partial\theta_d) l(\theta; y) = \nabla_\theta^T l(\theta, y), \qquad (A1.13)$$

say, with a corresponding definition of the random variable U.

A fuller notation would show explicitly that the score function depends on the particular choice of parameter as well as on the value of that parameter by writing $u^{(\theta)}(y, \theta)$. Then if we reparametrize by writing $\phi = \phi(\theta)$, a differentiable (1, 1) transformation, we have that

$$u^{(\phi)}(y, \phi) = (d\theta/d\phi) u^{(\theta)}(y, \theta),$$

where in the vector parameter case $d\theta/d\phi$ is replaced by $\partial(\theta)/\partial(\phi)$, the Jacobian matrix of the transformation.

We shall assume that a unique maximum likelihood estimate $\hat{\theta}$ exists inside the parameter space and thus produces a stationary value of the log likelihood satisfying

$$u(y, \hat{\theta}) = 0, \qquad (A1.14)$$

in general a set of d equations in d unknowns. We assume that either the equations have a unique solution or, if not, that the overall maximum is taken. We do not deal with applications in which the maximum is achieved or approached at the boundary of the parameter space.

It is easily shown that under suitable regularity conditions

$$E\{u(Y, \theta); \theta\} = 0, \qquad (A1.15)$$

the notation meaning that the expectation is evaluated at the same value of θ as is used to calculate u.

Now, again under suitable regularity conditions, the random variables

$$U^2 = \{u(Y, \theta)\}^2, \quad -\partial^2 l(\theta; Y)/\partial\theta^2 = -\partial u(Y, \theta)/\partial\theta \quad \text{(A1.16)}$$

have equal expectations, provided that the expectation is evaluated at the same value of θ as is used to define the random variables. The common value is called the expected or Fisher information and is denoted by $i(\theta)$. This is for a scalar parameter. For a vector parameter we define an information matrix $i(\theta)$ whose (s, t)th element is

$$
\begin{aligned}
i_{st}(\theta) &= E\{(\partial l(\theta; Y)/\partial\theta_s)(\partial l(\theta; Y)/\partial\theta_t)\} \\
&= E\{-\partial^2 l(\theta; Y)/\partial\theta_s\partial\theta_t\}.
\end{aligned}
\quad \text{(A1.17)}
$$

In the notation of (A1.13) we can write

$$i(\theta) = E\{-\nabla_\theta\nabla_\theta^T l(\theta, Y)\}.$$

It is useful also to define the observed information in the scalar case by

$$j(\theta) = -\partial^2 l(\theta; Y)/\partial\theta^2 \quad \text{(A.18)}$$

and in the vector case by $j(\theta) = -\nabla_\theta\nabla_\theta^T l(\theta, y)$. Nearly always we consider this at the maximum likelihood point, i.e. consider $j(\hat{\theta})$. The corresponding observed information matrix for a vector parameter is defined similarly as minus the matrix of second partial derivatives of the log likelihood.

These definitions of expected and observed information concern the information in the full vector y. Sometimes it is convenient to work with the information per observation, $\bar{i}(\theta) = i(\theta)/n$, $\bar{j}(\theta) = j(\theta)/n$.

Note that these definitions yield functions of θ but depend also on the choice of parameter and in a fuller notation we would write, for example, $i^{(\theta)}(\theta)$, the superscript denoting the defining parameter. In the scalar case if we change from θ to $\phi = \phi(\theta)$, the information becomes

$$i^{(\phi)}(\phi) = i^{(\theta)}(\theta)(d\theta/d\phi)^2 \quad \text{(A1.19)}$$

and in the vector parameter case the new information matrix is

$$\{\partial(\theta)/\partial(\phi)\}^T i^{(\theta)}(\theta)\{\partial(\theta)/\partial(\phi)\}, \quad \text{(A1.20)}$$

where $\partial(\theta)/\partial(\phi)$ is the Jacobian matrix of the transformation. These formulae follow immediately from the definition in terms of the covariance matrix of the score function. The same transformation law applies to the observed information only if this is evaluated at the maximum likelihood point. Note that the position of the maximum likelihood estimate is unchanged by reparametrization.

It follows on linearization of the equation (A1.14) and use of (A1.16) that when the amount of information is large $\hat{\theta} - \theta$ is asymptotically normally distributed with mean zero and variance $\{i(\theta)\}^{-1} \simeq \{j(\theta)\}^{-1}$ in the scalar case and multivariate normal with this covariance matrix in the vector parameter case.

Example A1.5 Normal-theory linear model
The log likelihood for the linear model of Section A1.2 under normal-theory assumptions is

$$- n \log \sigma - (y - x\beta)^{\mathrm{T}}(y - x\beta)/(2\sigma^2). \tag{A1.21}$$

The parameter is in general $\theta = (\beta, \sigma)$ but for the moment we regard σ as known. Then simple calculation shows that the score function is exactly (A1.4), the maximum likelihood estimate of β is the least squares estimate, and the expected and observed information matrices are equal, both being

$$(x^{\mathrm{T}}x)/\sigma^2. \tag{A1.22}$$

Further the asymptotic result for the distribution of the vector of maximum likelihood estimates holds exactly.

Example A1.6 Scalar exponential family
Suppose that a single observation y has the density

$$\exp\{s\phi - K(\phi)\}a_1(y), \tag{A1.23}$$

where s is a function of y and ϕ is a scalar parameter. The function $a_1(y)$ does not depend on ϕ and plays no direct role in what follows.

It is easily shown that the cumulant generating function of S under (A1.23) is

$$\log E(e^{Sp}) = K(\phi + p) - K(\phi), \tag{A1.24}$$

so that in particular the mean and variance of S are $K'(\phi)$ and $K''(\phi)$, the latter being also the information about ϕ in a single observation. The parameter ϕ is called the canonical parameter and s the canonical statistic. Other parametrizations can be used and one important choice is of

$$E(S; \phi) = K'(\phi) \qquad (A1.25)$$

which we call the moment parameter and denote by η. Note that the expected information functions for ϕ and for η are respectively

$$i^{(\phi)}(\phi) = K''(\phi), \qquad i^{(\eta)}(\eta) = i^{(\phi)}(\phi)(d\phi/d\eta)^2 = 1/K''(\phi). \qquad (A1.26)$$

Now suppose that instead of a single observation we have n responses represented by independent and identically distributed random variables with the density (A1.23). The log likelihood can be written

$$s\phi - nK(\phi), \qquad (A1.27)$$

where $s = \sum s_k$, s_k referring to the kth observation.

Now it follows directly from (A1.27) that the maximum likelihood estimate $\hat{\phi}$ satisfies

$$s = nK'(\hat{\phi}) = n\hat{\eta}, \qquad (A1.28)$$

so that the maximum likelihood estimate is obtained by equating the canonical statistic to its expectation. Further the expected and observed information functions for ϕ are both equal to $nK''(\phi)$.

Finally it is particularly important that these results require only that the log likelihood has the structure (A1.27) for a scalar s and are not restricted to independent and identically distributed random variables.

Example A1.7 Independent and identically distributed binary trials
An especially important special case of Example A1.6 arises when the single observations are binary with probability of 'success' θ, say. For $y = 0, 1$ the probability can be written

$$\exp[y\log\theta + (1 - y)\log(1 - \theta)]$$
$$= \exp[y\log\{\theta/(1 - \theta)\} + \log(1 - \theta)], \qquad (A1.29)$$

which is of the required form with canonical statistic y, canonical parameter $\log[\theta/(1 - \theta)]$ and moment parameter equal to the expected value of Y, namely θ. To complete the parallel with (A1.23) we express (A1.29) in terms of the canonical parameter, obtaining

$$\exp\{y\phi - \log(1 + e^\phi)\} \qquad (A1.30)$$

and the various general properties of Example A1.6 can be verified with $K_1(\phi) = \log(1 + e^\phi)$. When we deal with n independent and identically distributed binary trials s is the total number of successes.

Example A1.8 Full exponential family

Suppose that the likelihood for the full vector y has the form

$$\exp\{s_1\phi_1 + \cdots + s_d\phi_d - nK(\phi)\}a(y). \qquad (A1.31)$$

Here $s_1, \ldots, s_d$ are functions of y and $\phi_1, \ldots, \phi_d$ are independent parameters, the key point being that the dimension of the minimal sufficient statistic is the same as that of the parameter. We call (A1.31) the (d, d) exponential family. Note that while the likelihood may arise from independent and identically distributed observations this is not necessary, only the structure of the likelihood being relevant. The likelihood for the linear logistic regression for binary data is of this more general form.

The results and definitions of Example A1.6 generalize. Thus

$$E(S_t; \phi) = n\,\partial K(\phi)/\partial\phi_t = n\eta_t, \qquad (A1.32)$$

defining the moment parameter η. That is, $\nabla_\phi K(\phi) = \eta$. The maximum likelihood estimate satisfies

$$s_t = n\hat{\eta}_t \qquad (t = 1, \ldots, d), \qquad (A1.33)$$

and the information matrix for the canonical parameter ϕ is the Hessian matrix of $nK(\phi)$, i.e. is $n\nabla\nabla^T(\phi)$. If the observations are represented by n independent and identically distributed random variables, then the function K is that for a single observation, but the results are not restricted to that case.

Example A1.9 Curved exponential family

A rather general family of likelihoods is obtained by starting with the (d, d) exponential family and then supposing that the parameters ϕ_s are expressed, in general non-linearly, in terms of $d_\beta < d$ unknown parameters β; equivalently the moment parameters η_s are specified in terms of β. We call this the (d, d_β) curved exponential family. It is a consequence of the non-linearity that the minimal sufficient statistic remains of dimension d.

The log likelihood is

$$\sum s_t\phi_t(\beta) - nK\{\phi_1(\beta), \ldots, \phi_d(\beta)\}, \qquad (A1.34)$$

where the moment parameters are still defined by

$$\eta_t = E(S_t; \phi) = n\partial K/\partial \phi_t. \qquad (A1.35)$$

It follows on differentiation with respect to β_u, a component of β, that the maximum likelihood estimates satisfy

$$\sum (s_t - n\hat\eta_t)\partial \phi_t/\partial \beta_u = 0, \qquad (\nabla_\beta \phi^T)(s - n\hat\eta) = 0. \qquad (A1.36)$$

Note that this is an equation in $\hat\beta$ satisfied at the maximum likelihood point and that in general $\hat\beta$ enters all parts of the equation. Now $\eta = \nabla_\phi K$, so that

$$\nabla_\beta \eta^T = (\nabla_\beta \phi^T)(\nabla_\phi \nabla_\phi^T K),$$

via which we can replace (A1.36) by

$$(\nabla_\beta \eta^T)(\nabla_\phi \nabla_\phi^T K)^{-1}(s - n\hat\eta) = 0. \qquad (A1.37)$$

Now this has the form of a generalized least squares equation (A1.10) in which the 'design' matrix x is replaced by $\nabla_\beta \eta^T$ and in which the covariance matrix V is replaced by $\nabla_\phi \nabla_\phi^T K$.

The information matrix for $\hat\beta$ has (u, v) element

$$\sum (\partial \phi_r/\partial \beta_u)(\partial \phi_t/\partial \beta_v)i_{rt} \qquad (A1.38)$$

or in matrix form is

$$\begin{aligned} n(\nabla_\beta \phi^T)(\nabla_\phi \nabla_\phi^T K)(\nabla_\beta \phi^T)^T \\ = n(\nabla_\beta \eta^T)(\nabla_\phi \nabla_\phi^T K)^{-1}(\nabla_\beta \eta^T)^T. \end{aligned} \qquad (A1.39)$$

This completes the close formal and conceptual connection between maximum likelihood estimation and generalized least squares estimation. That is, if we repeatedly apply generalized least squares redetermining the weights and the 'design' matrix as functions of the parameters until convergence is achieved, then the maximum likelihood estimates and their covariance matrix are recovered. That is, maximum likelihood estimation is generalized least squares with appropriately consistent weights and design matrix. Geometrically maximum likelihood estimation corresponds to projection in a metric determined locally in the neighbourhood of the maximum likelihood point.

Example A1.10 Generalized linear model
While in many ways Example A1.9 specifies a natural fairly general framework for seeing the relation between maximum likelihood estimation and least squares, from a more applied point of view it is

helpful to consider rather less general settings. This involves a number of rather different specializations. We start with the (d, d) model and suppose first that the canonical statistics are mutually independent, usually because they arise from independent observations. Then $nK(\phi_1, \ldots, \phi_d)$ is a sum

$$\sum K_t(\phi_t)$$

and often the K_t are of similar form. The additive structure implies that the matrix $\nabla_\phi \nabla_\phi^T K$ is diagonal.

Secondly it may be reasonable to specify a linear relation for each ϕ_t, for example that

$$\phi_t = h(\sum x_{tu} \beta_u) \tag{A1.40}$$

or equivalently that

$$\eta_t = g(\sum x_{tu} \beta_u). \tag{A1.41}$$

Here h and g are known functions and the x's are explanatory variables. Quite often it is helpful to replace (A1.40) by

$$\phi_t = \beta_0 h(\sum x_{tu} \beta_u). \tag{A1.42}$$

Both these models are natural and fruitful generalizations of linear regression but unless the function h is linear there is no great theoretical simplification over Example A1.11. If, however, h is linear we have an important special case which we treat separately.

Example A1.11 Canonical linear model
We start with the (d, d) exponential family and suppose that the canonical statistics are independent and that the canonical parameters have linear representations in terms of a smaller number d_β of unknown parameters so that

$$\phi_t = \sum x_{tu} \beta_u. \tag{A1.43}$$

On substitution into the log likelihood (A1.34) it is clear that we obtain a reduced (d, d) family with a new set of canonical statistics and that we can take the β as canonical parameters. The discussion of Example A1.8 applies with log likelihood

$$\sum (\sum s_t x_{tu}) \beta_u - \sum K_t(\sum x_{tv} \beta_v). \tag{A1.44}$$

Binary logistic regression is an important special case. Note that the treatment of probit linear regression would require a special case of

Example A1.10. From the point of view of maximum likelihood theory in general there is not much difference between the two cases; however, in the second the data enter only through d_β rather than through d functions; also in the (d_β, d_β) family there is the possibility of 'exact' inference for suitable target parameters.

A1.5 Techniques for inference

We now turn to methods based directly on the log likelihood function for finding confidence limits for a parameter of interest. There are broadly three such methods, each with relatively minor variants. If there is a single parameter θ with no nuisance parameters, the methods are as follows.

First in the likelihood ratio method we take as confidence region all those θ giving log likelihood sufficiently close to the maximum. More specifically we test the null hypothesis $\theta = \theta^{(0)}$ by the likelihood ratio statistic

$$w(\theta^{(0)}) = 2\{l(\hat\theta) - l(\theta^{(0)})\}, \tag{A1.45}$$

this having asymptotically a chi-squared distribution with d_θ degrees of freedom. This leads to the $1 - \alpha$ level confidence region

$$\{\theta; w(\theta) \leqslant \chi^2_{d_\theta, \alpha}\}, \tag{A1.46}$$

where $\chi^2_{d_\theta, \alpha}$ is the upper α point of the standard chi-squared distribution with d_θ degrees of freedom. In the special case $d_\theta = 1$ this is equivalent to treating $\text{sgn}(\hat\theta - \theta)\sqrt{w(\theta)}$ as a standard normal variable.

The second possibility is to use the score function directly, i.e. to take as confidence region the set of all θ with sufficiently small values of the gradient of the log likelihood. More precisely, we test the null hypothesis $\theta = \theta^{(0)}$ by the statistic

$$w_u(\theta^{(0)}) = u^{\mathrm{T}}(\theta^{(0)})i^{-1}(\theta^{(0)})u(\theta^{(0)}), \tag{A1.47}$$

this again having a chi-squared distribution with d_θ degrees of freedom. A confidence region for θ is formed analogously to (A1.46) by

$$\{\theta; w_u(\theta) \leqslant \chi^2_{d_\theta, \alpha}\}. \tag{A1.48}$$

If θ is a scalar, $d_\theta = 1$, $u(\theta^{(0)})/\sqrt{i(\theta^{(0)})}$ has a standard normal distribution. In the expressions (A1.47) and (A1.48) it is possible to replace $i(\theta^{(0)})$ by $j(\theta^{(0)})$.

The third possibility is to work with the maximum likelihood estimate $\hat{\theta}$, basing a test of $\theta = \theta^{(0)}$ on the difference $\hat{\theta} - \theta^{(0)}$ via the statistic

$$w_e(\theta^{(0)}) = (\hat{\theta} - \theta^{(0)})^{\mathrm{T}} i(\theta^{(0)})(\hat{\theta} - \theta^{(0)}). \qquad (A1.49)$$

Confidence regions for θ are formed as above by

$$\{\theta; w_e(\theta) \leqslant \chi^2_{d_\theta, \alpha}\}. \qquad (A1.50)$$

In this $i(\theta^{(0)})$ can be replaced by $i(\hat{\theta})$, $j(\theta^{(0)})$ or $j(\hat{\theta})$. If θ is a scalar we treat $(\hat{\theta} - \theta^{(0)})\sqrt{i(\theta^{(0)})}$ as a standard normal variable. The simplest confidence intervals come from using information evaluated at $\hat{\theta}$, from which we obtain limits of the form

$$\hat{\theta} \pm k^*_{\frac{1}{2}\alpha}\sqrt{i(\hat{\theta})} \qquad (A1.51)$$

or the corresponding form with observed information instead of expected information, $k^*_{\frac{1}{2}\alpha}$ being the one-sided $\frac{1}{2}\alpha$ point of the standard normal distribution.

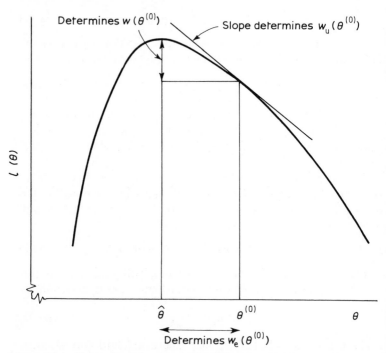

Figure A1.1. *Tests of null hypothesis $\theta = \theta^{(0)}$ via log likelihood function.*

Figure A1.1 illustrates these procedures. If the log likelihood is quadratic with curvature equal to its expected value all three procedures agree exactly. In the majority of applications the log likelihood will be nearly quadratic and the answers from these procedures will be in practical accord with one another. This is not always the case, however, and it is therefore important to consider the relative merits of the different procedures.

First there are some qualitative considerations.

The desirability that inference about a parameter θ should be essentially unchanged by transforming from θ to $h(\theta)$ points against the use of w_e or at least suggests that if maximum likelihood estimates are used directly an appropriate choice of parametrization has to be made.

The need to preserve sensible behaviour should there be multiple maxima points against the score statistic w_u. On the other hand the score statistic may have substantial computational advantages, especially when a fairly complicated model is to be amplified in various ways in order to test its adequacy. For then the only full fitting needed is under the null model.

The direct use of a likelihood ratio statistic, w, preserves invariance and is especially satisfactory in ensuring qualitatively sensible confidence regions whenever these are naturally asymmetric or certain values are logically impossible or when the likelihood approaches a supremum at infinity.

In summary then the direct use of w seems the safest 'routine' approach to new problems and leads to a direct presentation of conclusions via a profile log likelihood, although in simple cases a standard error attached to a maximum likelihood estimate is very appealing.

There are more elaborate considerations based on approximate conditioning that also point to the use of w.

The discussion thus far has concerned situations with no nuisance parameters. At one level the results generalize immediately. Suppose that $\theta = (\psi, \lambda)$, where ψ is the parameter of interest, often but not necessarily a scalar, and λ is a nuisance parameter. Then the simplest definition of a log likelihood function for ψ is the profile log likelihood

$$l(\psi, \hat{\lambda}_\psi), \tag{A1.52}$$

where $\hat{\lambda}_\psi$ is the maximum likelihood estimate of λ for given ψ. The idea is then to treat the profile log likelihood exactly as if it were an ordinary log likelihood for a parameter of dimensionality d_ψ. Thus the null hypothesis $\psi = \psi^{(0)}$ is tested via

$$w(\psi^{(0)}) = 2\{l(\hat{\psi}, \hat{\lambda}) - l(\psi^{(0)}, \hat{\lambda}_{\psi^{(0)}})\}. \tag{A1.53}$$

Note that this is exactly in the same spirit as the technique (A1.9) used for the normal-theory linear model. Confidence regions for ψ are formed from all values of ψ giving profile log likelihood sufficiently close to the maximum using (A1.46) now with d_ψ degrees of freedom.

In the terminology of the computer program GLIM the formal statistic (A1.53) for comparing $\psi = \psi^{(0)}$ with a saturated model containing the maximum number of nuisance parameters for the problem is called the deviance. Often, and especially with binary data, the deviance itself is of little interest and in particular does not have the formal chi-squared distribution because the number of nuisance parameters in the saturated model is large. The relevant quantity for use in (A1.53) is the difference between the deviance attached to $(\psi^{(0)}, \lambda)$ and that for (ψ, λ), where λ has a suitably modest number of components.

Many of the qualitative advantages of the procedure without nuisance parameters are retained, but there can be serious difficulties if the dimensionality of the parameters is comparable with the number of observations.

The score statistic w_u and the maximum likelihood estimate statistic w_e generalize in essentially the same way as w. Thus for w_e we extract the covariance matrix of $\hat{\psi}$, say $i^{\psi\psi}$, from the appropriate section of the inverse information matrix, and then to test $\psi = \psi^{(0)}$ we calculate

$$w_e(\psi^{(0)}) = (\hat{\psi} - \psi^{(0)})^{\mathrm{T}} (i^{\psi\psi})^{-1} (\hat{\psi} - \psi^{(0)}), \tag{A1.54}$$

where in the information matrix λ can be estimated in any consistent fashion, for example by $\hat{\lambda}$. Note that by a standard formula for the inverse of a partitioned matrix

$$(i^{\psi\psi})^{-1} = i_{\psi\psi} - i_{\psi\lambda} i_{\lambda\lambda}^{-1} i_{\psi\lambda}^{\mathrm{T}}. \tag{A1.55}$$

Finally the generalization of the score involves extracting the section u_ψ of the score function referring to ψ, estimating λ by $\hat{\lambda}_\psi$, and considering the quadratic form

$$w_u(\psi^{(0)}) = u_\psi^{\mathrm{T}} i^{\psi\psi} u_\psi, \tag{A1.56}$$

the primary advantage of this being that it involves maximum likelihood fitting only under the null hypothesis.

Note that when, as is frequently the case, the parameter ψ of interest is a scalar, the use of w_e with precision estimated via the observed or expected information matrix at the maximum likelihood point yields confidence regions summarized by a point estimate and an approximate standard error.

Many of the qualitative arguments in favour of direct use of the likelihood ratio statistic, w, i.e. in favour of use of the profile log likelihood, remain, but the method may break down badly if the number of nuisance parameters is large and presumably therefore is capable of some improvement whenever nuisance parameters are present. This is discussed briefly in the next section.

A1.6 Some warnings

In the previous section we have outlined a general procedure based on profile log likelihood for obtaining confidence regions for a parameter of interest, usually in the presence of nuisance parameters. There are some potential difficulties which we now list briefly.

First maximum likelihood estimates must by definition be points in the parameter space for the problem. If the parameter space is non-standard and particularly if the maximum is achieved on the boundary of the parameter space the usual results will not apply. A rather similar situation arises in so-called non-regular problems, in which the maximum likelihood estimates are related to extreme order statistics and do not satisfy the usual estimating equation.

A quite different difficulty arises if there are many nuisance parameters, notably when the number of nuisance parameters is comparable with the number of observations. Then the maximum likelihood estimate may be far from the true value. Two ways round this are first by eschewing models with large numbers of nuisance parameters, for example by use of empirical Bayes formulations, or secondly by some exact or approximate modification of the likelihood, for example by conditioning.

A1.7 Some further developments

We mention briefly two further developments of the above discussion.

First if we are giving a central role to the likelihood ratio statistic w

it is natural to aim to improve on the chi-squared distribution which is used as a convenient approximation to its distribution. It turns out that this is surprisingly simple. If w is exactly distributed as chi-squared with d degrees of freedom, then $E(w) = d$. Suppose that in fact

$$E(w) = d(1 + b/n), \qquad (A1.57)$$

where b is either known or can be estimated. Then the statistic

$$w' = w/(1 + b/n) \qquad (A1.58)$$

has 'improved' expectation but remarkably the whole distribution is made closer to the chi-squared form. The factor $1 + b/n$ is called a Bartlett adjustment.

The second general point concerns modifications of profile log likelihood to improve performance, especially when there are many nuisance parameters. While the development of such modifications is the subject of active work at the time of writing, no widely agreed easily implemented form is available.

A1.8 Conditioning

The previous sections of this Appendix have been largely concerned with maximum likelihood theory whose theoretical justification is approximate. That is, it relies largely on local linearization and on asymptotic normality arising from the central limit theorem. When applying the results it needs to be considered whether the amount of information available is large enough for adequate approximations to be achieved.

We now turn briefly to the role of conditioning in achieving 'exact' results in some important, but nevertheless very special, cases. Of course 'exact' here means 'without mathematical approximation within the framework of an assumed model'. Because all such models are at best good approximations to the real scientific situations under study, 'exactness' in the present sense may seem of limited immediate importance. On the other hand, it is essential to know when the mathematical approximation involved in using asymptotic theory is in some sense small and from this point of view the availability of some 'exact' answers is of considerable value as providing a reference standard for assessing other procedures, although of course it also has immediate application in simple cases.

Conditioning, i.e. the use for inference of sampling distributions conditioned on suitable aspects of the data, arises in at least two broad contexts related in a loose way. These are first to ensure that the distributions used for inference are relevant to the interpretation of the particular data under analysis, and second to construct distributions useful for inference that do not involve unknown nuisance parameters.

As to the first aspect, the key point is that the probability calculations involved in confidence intervals and significance tests provide explanations of these procedures in terms of frequencies in hypothetical repetitions, i.e. in terms of what would happen if such and such a procedure were applied again and again. For such an explanation to be relevant to a particular application, it is necessary that the hypothetical set of repetitions contemplated should be relevant to the application in question and this is achieved by conditioning on appropriate features. The precise formulation and implementation of this can be quite difficult, in part because overconditioning can lead to appreciable loss of sensitivity. The general idea is, however, compelling if measures of uncertainty are to be based on a frequency interpretation of probability and if statistical procedures are to be calibrated conceptually in terms of what happens when they are used.

The second aspect of conditioning is in many ways more easily formalized. In the context of the (d, d) exponential family if the problem can be set up so that interest is focused on one canonical parameter (or on a ratio of canonical parameters), then it can be shown that the only way to obtain a distribution for assessing significance that does not depend on the nuisance parameters is by conditioning on the values of the remaining canonical statistics, i.e. those not associated with the parameter of interest. Because of the relation between canonical statistics and moment parameters this amounts to the result that for inference about the canonical parameter ϕ_1 we should condition on the maximum likelihood estimates $\hat{\eta}_2, \ldots, \hat{\eta}_d$. There are in the text a number of examples of this to the linear logistic model.

This use of conditioning rests on a clear-cut theorem about the ways of obtaining distributions free of nuisance parameters. The only limitation on use, given the broad approach, lies in the possibility that in very small discrete problems there may be a loss from overconditioning.

A1.9 Bayesian approach

The broad approach of the book has been non-Bayesian, in the sense that we have not routinely suggested the tackling of problems via the formulation of prior distributions. We conclude the Appendix by a brief comment on this issue.

Bayesian arguments can be used in three rather different ways. The first, really quite uncontentious, arises when there are a number of parameters measuring essentially the same quantity in a range of related circumstances. The parameters may be ones of direct interest, for example the difference between two treatments as obtaining in a range of similar but different environments, or may be nuisance parameters specifying individual subject effects in a treatment comparison evaluated on a collection of subjects. Then it may be helpful to treat the parameters in question as having a frequency distribution, usually one to be estimated empirically from the data under analysis. This gives rise to the name empirical Bayes.

The object in the second case is economy of parametrization and in the first to 'borrow strength' in the estimation of an effect in one environment by combining the estimate in that environment with the results in other environments. For brief discussions of this see Section 3.3.

The second use of Bayesian arguments is as an alternative to confidence interval estimation, deriving posterior intervals by assuming flat priors for unknown parameters. That is, the prior distribution represents rough initial ignorance so that the posterior distribution summarizes the information from the data. Many of the confidence interval procedures in the book can be derived exactly or approximately by that route. We have chosen not to follow that path largely because of the conceptual and technical difficulties connected with 'flat' priors.

The third role of Bayesian arguments, and the one that has received most attention in recent discussions, is as a route for inserting into the analysis quantitative information about the investigator's judgement or opinion about the matter under discussion. It is assumed that this can be done by the investigator specifying a prior subjective probability distribution for, say, the parameter in question, based on all considerations other than the data under analysis. We certainly do not question the value, relevance and indeed necessity of judgement in the interpretation of data, very particularly in the formulation of questions for study. The usefulness of judgement formulated in the above way and of its merging with the information from the data must depend on the context. In most questions of the careful assessment and reporting of scientific conclusions we think that the information about the values of parameters from the data and that from prior information should so far as feasible be kept separate, and that is our reason for not using subjective prior distributions in the present book.

For more detailed discussion of the topics in this Appendix, see the books by Cox and Hinkley (1974) and McCullagh and Nelder (1983).

Choice of explanatory variables in multiple regression

A2.1 Introduction

In Section 2.1 we set out procedures for analysing a general linear logistic regression in which dependence of a probability of success on explanatory variables is expressed via a linear relation on the logistic scale. Implementation of these methods is straightforward once a set of explanatory variables is chosen for inclusion. In applications, however, a crucial aspect is precisely that choice, it being fairly rarely the case that, for example, theoretical considerations indicate unambiguously the equation to be fitted. The points at issue are similar to those arising with 'ordinary normal theory' empirical multiple regression based on the method of least squares and indeed in other forms of empirical linear regression in generalized linear models. In the present Appendix we discuss these issues in fairly broad terms.

One special aspect of the normal-theory case that does affect the strategy of tackling the analysis of highly balanced sets of data is that exact or nearly exact orthogonality in normal theory implies that estimates of certain parameters are unaffected by the inclusion or exclusion of some other parameters. This makes it feasible to begin the analysis of a balanced design by inspection of a 'full' analysis of variance in which possibly large numbers of main effects and interactions are included. In linear logistic regression however a balanced design leads to only approximate orthogonality of the estimated parameters and it is not always possible to see immediately the precise effect of such inclusion or exclusion. For this reason it is commonly sensible to begin with some relatively simple model and then to examine the need to amplify or indeed simplify the initial model. The criteria for the choice of that starting model as well as for modifying the model in the light of the data become of more pressing concern.

Sections A2.2 and A2.3 deal with type and formation of explanatory variables rather than with the strategy for choice of explanatory variables but these ideas are nevertheless important in the analysis and interpretation of regression models. Also much of the material in Sections A2.4 and A2.5 is not specific to binary data but is given here for completeness of discussion. Reference is made to examples discussed in Chapter 2 of this book.

A2.2 Types of explanatory variable

It is convenient to classify potential explanatory variables in several different ways.

First for purposes of interpretation we may classify explanatory variables as in Section 2.8, namely as

1. treatment or quasi-treatment variables representing aspects which can in principle at least be manipulated (Example 2.10);
2. intrinsic variables measuring aspects characterizing an individual under study or the environment in which the study on an individual is carried out, for example age, socio-economic class (Example 2.18);
3. non-specific variables characterizing broad groupings of individuals; often such groupings are described by names such as blocks, strata and so on (Example 2.16).

The object of study is normally the assessment of the effect of treatments and of possible interaction of treatment effects with variables of type (2) or (3) (Example 2.17).

Of course this division into three types depends on the context and may not be clear-cut. Especially in observational studies some intrinsic variables, such as socio-economic class of individuals, may be surrogates for other more specific properties, such as educational background, and wherever possible more specific variables should, of course, be used.

In a randomized experiment, treatment variables are randomized and intrinsic variables are those measurements made on the individuals before randomization.

In an observational study, the treatments are typically aspects that ideally have been investigated via a randomized experiment, but which in fact were determined in a way outside the investigator's control. Thus in a study of the effect of alcohol consumption during pregnancy on some feature of the infant, randomization is obviously not feasible with human subjects. Treatment, perhaps better called quasi-treatment, variables are thus measures of alcohol consumption and other matters, such as diet, necessary to define the treatment effect under study, whereas intrinsic variables are mother's age and parity, socio-economic class, etc. If the study were replicated in a number of centres, centres would form a non-specific variable.

A second classification of explanatory variables, relevant in analytical formulation, is by their mathematical structure, according to whether they take

1. a number of qualitatively different levels, such as one of a number of regions of residence;
2. a number of ordered levels, such as the description of the severity of some condition as slight, moderate, severe and very severe;
3. values specified by a reasonably well-defined quantitative scale.

A third classification is into

1. directly measured variables;
2. derived variables, by which we mean both composite variables obtained by taking combinations of measurements or variables such as squares and products of more directly observed quantities.

A2.3 Formation of explanatory variables

In some situations explanatory variables may be entered into a multiple regression equation either in exactly the form in which they are measured or after rescaling; a simple change of units to make all variables have approximately the same standard deviation in the data, and in some cases a change of origin to produce means that are not too large may help avoid numerical instability (Example 2.11). For essentially positive variables a log transformation may be wise (Examples 2.10, 2.11). Care is, however, needed with variables that have a very wide range, especially where very non-linear effects are likely. Thus if in a clinical study age at entry ranged from 60 to 70 years, direct introduction of, say, age $- 65$ as a quantitative variable would be reasonable, and non-linearity could, if necessary, be checked via a squared term. But if age ranged from 20 to 80 years some grouping of age into a fairly small number of groups and their treatment initially as qualitatively different, as explained below, would protect against strong non-linearity. Again for alcohol con-

p. 66.

sumption in litres per week it would usually be better to work initially with none, slight, moderate, heavy rather than directly with the quantitative measurement; later analysis could refine the initially arbitrary subdivision, if that seemed likely to be fruitful.

For qualitative variables at l levels, the construction of $l-1$ explanatory variables will be needed if the main effect of such a variable is to be represented without prior constraint (Example 2.12). The method of construction is in one sense arbitrary so long as we make the $l-1$ variables linearly independent but the following considerations are helpful.

1. The marginal frequencies of the different levels should be inspected, in particular to avoid giving prominence to levels that occur with very low frequency.
2. If there is a level, say 1, with a very low frequency and its possible merging with another level, say 2, appears possibly sensible, it will be useful to define one variable, say $x_1 = 1$ (level 1), -1 (level 2), 0 (all other levels), so that the resulting estimated parameter provides a test of the reasonableness of the proposed merging. In defining the other $l-2$ variables, the two levels 1 and 2 can then be treated identically.
3. If one level, say 1, is a control, or other natural reference level, or occurs with especially high frequency, it may be sensible to define all the xs relative to 1, i.e. to define $x_1,\ldots,x_{l-1}$ by $x_j = 1$ (level $j+1$), -1 (level 1), 0 (otherwise).
4. If the levels are ordered and are of very roughly equal frequency, it may be sensible to define x's via the standard orthogonal polynomials (Pearson and Hartley, 1966, Table 47) for l equally spaced points, using thus for three levels $-1,0,1$; $1,-2,1$ to define respectively x_1, x_2.
5. There should be some rough check that the xs are not defined so as to be nearly linearly dependent.
6. Any special arguments indicating contrasts that are likely to be particularly important should, of course, be used in defining the xs.

Interactions are normally best studied in this context by defining products of the xs defining the main effects in question. In exploratory work, the principle that large main effects are on the whole more likely to generate appreciable interactions than small main effects is often helpful. Thus if two qualitative variables have

l_1 and l_2 levels respectively, leading to the definition of $l_1 - 1$ and $l_2 - 1$ explanatory variables, the set of all products of these variables defines the two-factor interaction with $(l_1 - 1)$ $(l_2 - 1)$ degrees of freedom. If it is required to extract a few degrees of freedom from the interaction this can be done via products of component x's with especially strong interpretations or, failing that, via components that happen to be large.

In some applications to calculate *a priori* combinations of the explanatory variables may be useful. For example: near orthogonality may be achieved by replacing diastolic and systolic blood pressure by the sum and difference of the logs of the two measurements; if a particular feature has been measured in several different ways a composite score may initially be tried. In these cases it will often be wise to test from the data whether the indicated combination appears to have sacrificed information about the response variable under study.

A2.4 Small numbers of explanatory variables

In some applications there may be a reasonably small number of explanatory variables, corresponding to say at most five or six parameters. Unless a treatment effect of primary interest is substantially confounded with variables of no direct interest, there seems little point in trying to simplify the resulting equation by omitting explanatory variables merely on grounds of statistical insignificance; it may be a useful quick check on the potential for improving precision of treatment effects to compare the standard error under the full fit with that achieved by omission of *all* other variables.

The model with all explanatory variables in linear form may be augmented by adding non-linear functions, e.g. squares of quantitative variables, and interaction terms (Example 2.11). A simple strategy is to begin by adding such terms one at a time, concentrating on interactions of the treatment effects of primary interest with intrinsic and non-specific explanatory variables and on possible non-linearity of response to important quantitative variables.

In judging statistical significance it is important to make allowance whenever the largest from among many possible contrasts is chosen for interpretation. One way to do this, when a variable is chosen for inclusion out of a block of variables, is to examine the change in 2 log (maximized likelihood) when all the variables in the block are fitted; if this is not statistically significant at an interesting level, there is a danger that the variable selected is an artefact.

A2.5 Large numbers of explanatory variables

A much more difficult situation arises if there are so many potential explanatory variables that some reduction from the full fit is essential either to achieve understandable interpretation or reasonable precision in the primary comparisons. Now many computer packages contain automatic algorithms for variable selection. We strongly recommend against reliance on these algorithms, except occasionally in the very restricted context set out below. This is because the choices they force are often of a very arbitrary character and are often not the most appropriate for the purpose either of prediction or of interpretation; to end with one set of variables when there are other quite different choices having virtually as good a fit to the data invites misinterpretation.

We suggest a procedure along broadly the following lines.

1. List those explanatory variables which it is essential to include either because they are treatment variables of primary concern, or because it is known from previous studies that they are important.
2. Consider whether certain subsets of variables (e.g. measurements of the same feature) should be treated separately or whether some preliminary reduction across subsets, such as by the formation of totals, might be fruitful.
3. Check for the influence of other variables, at first one at a time as in Section A2.4, or by cautious use of a computerized selection algorithm.
4. Iterate this procedure, i.e. repeat both phases with the initial variables, those from the initial list supplemented by and/or replaced by other variables found empirically from the data.
5. When one or more apparently adequate fits have been obtained check for the addition of further variables, interactions and so on as outlined in Section A2.4.

If, as is likely, there are different choices in the later phases that give adequate fits it is important, as far as is feasible, to give *all* fits consistent with the data, making any choice between alternative choices on subject-matter grounds.

A crucial aspect is the behaviour under alternative models of the aspects of primary interest, i.e. parameters representing treatment effects and their potentially important interactions with intrinsic variables. If these are reasonably stable, choice of other aspects of the model is probably not of critical importance.

Note especially that important variables, especially those representing treatments, should not be excluded solely because the corresponding estimates are insignificant statistically; their estimation is likely to be of direct interest and the inclusion of estimates and standard errors is in any case likely to be essential in a final report on the data, if only to allow comparison with subsequent related studies.

Where there are several rather similar sets of data for analysis it will usually be wise to use the same explanatory variables for all sets. Thus if the number of such variables is large a cautious procedure is to aim to choose explanatory variables separately for each set and then to re-analyse using the set of *all* variables so chosen, before possibly attempting some common reduction. Incautious use of automatic selection algorithms is quite likely in these contexts to throw up different choices of variables in the different sets, with consequent dangers of misinterpretation.

A final note of caution concerns the interpretation of significance tests and confidence intervals when a complex sequence of data-dependent choices has been made. If there is really very little or no explanatory power in a block of variables and one or two are selected on the basis of the data as showing the largest apparent effect, there is a clear possibility of considerable exaggeration of significance; this stresses the need for some protection from global tests of blocks of parameters.

We do not think it feasible to specify probability properties of complex sequences of data-dependent choices. Note, however, that if *all* sufficiently simple equations consistent with the data at a specified standard significance level are listed, any 'correct' such specification will be included with the specified confidence coefficient.

We consider, however, that by following the broad guidelines above and concentrating on the treatment effects of primary concern, these puzzling difficulties are to a large extent bypassed. If all that is required is a well-fitting empirical equation and no interest attaches to individual effects, again questions of the significance of individual terms are essentially irrelevant, although we regard such totally empirical prediction equations as of rather limited interest.

APPENDIX 3

Review of computational aspects

A3.1 Introduction

Computer packages are readily available for carrying out most standard statistical calculations. Many packages are being written for use on microcomputers and even the larger packages, originally developed for main frame systems, now have versions for mini- or microcomputers.

Considerations involved in the choice of a package depend upon the nature of the statistical analysis and on the extent of the data to be analysed. What is best for exploratory work may not be the most suitable for simple use by occasional users. Some of the desiderata to be borne in mind are: numerical stability of the underlying algorithms and, possibly, speed of calculation; flexibility of data manipulation and the ability to handle large sets of data; well-labelled output and flexibility in controlling the amount and the nature of the output; the ability to use output from one analysis as input for further analysis; flexibility in analysis, allowing e.g. fitting non-standard models, control of selection of variables, deletion of lines of data, coping with missing values, computation of non-standard residuals or diagnostics; quality of graphical output; availability of interactive use; ease of use by the inexperienced or occasional user.

A3.2 Specific programs

It is not feasible to mention all programs which are currently available and suitable for analysis of binary data. Larger programs which are widely available include GLIM, GENSTAT, BMDP, SAS and SPSSX.

GLIM (Generalized Linear Interactive Modelling) was designed from the start to allow exploratory analysis GLIM embodies greater flexibility than most programs, but has less well-labelled output. The GLIM newsletter provides a useful source of GLIM macros.

GLIMPSE (Generalized Linear Interactive Modelling with PROLOG and Statistical Expertise) is a knowledge-based front-end (see Section A3.3) for GLIM, designed to make the program easier to use by non-experts by incorporating statistical expertise into the system and allowing the user to give instructions to the system at a higher level than required by most programs. It incorporates extensive model-checking techniques.

GENSTAT (General Statistical program) bears many similarities to GLIM, both having been developed through Rothamsted Experimental Station. It is a larger program with better data-handling and graphics facilities. Originally it was written for batch operation but it can also be run interactively.

BMDP (Biomedical Programs) is a library of integrated programs. Program LR fits logistic regression, 3R is a non-linear regression program which allows the user freedom to fit any specified model, and 4F will tabulate data or fit loglinear models. Certain items of output may be used as input into other programs.

SAS (Statistical Analysis System) is a larger system of integrated programs. Logistic regression can be fitted by LOGIST or CATMOD procedures, the latter permitting also more general transformations of the response variable. The procedure FREQ permits analysis of several 2×2 tables based on the Mantel–Haenszel estimate.

SPSSX (Statistical Package for the Social Sciences) is a system developed originally for survey analysis. It is simple to use, but generates a large amount of output. The LOGLINEAR procedure allows the fitting of loglinear and logistic models.

A3.3 Expert systems

There has been much recent discussion of expert systems for statistical analysis, e.g. GLIMPSE (Section A3.2). These provide strategical advice on how to use statistical methods, what questions to ask in deciding on a method of analysis, how to interpret answers, and so on.

The term expert system is a misnomer, at least with respect to its use in general artificial intelligence work. There it is supposed that there is a definitively correct answer in each situation. The object is to enable non-expert users to come close to that answer. Usually the system is calibrated by observing the performance of the 'expert'. For statistical analyses of realistic complexity it is more a matter of providing general advice on how to approach analyses, especially those involving relatively complicated methods.

We make a few general points about such systems. First there will normally be two components, a knowledge engine and a statistical engine. The latter may well be one of the packages, such as GLIM, described in Section A3.2.

In assessing the knowledge engine, the following points are relevant.

1. How authoritarian is it? Does it give advice and offer suggestions, e.g. on whether to omit suspect observations, or does it determine what to do in accordance with rigid rules?
2. How flexible is it, in for example allowing the user to proceed directly with straightforward parts of the analysis?
3. Is there a lexicon to explain unfamiliar technical terms and also information available on request to justify advice offered?
4. Does it allow easy passage in any order between various phases of analysis?

A related aspect concerns the type of user for whom the system is designed This may be

1. a user with little experience of quantitative methods of analysis;
2. a research worker who is an occasional user of advanced statistical methods;
3. a statistician who is inexperienced, either with the particular package in question or with the 'practical' aspects of the methods of analysis.

Guidance to the first type of user over simple graphical and tabular methods could be of much value. The most fruitful applications are, however, likely to be for the other types of user. For the theoretical statistician the interest lies in the need to clarify the strategical issues involved. Such issues may be encapsulated in computer systems or, even, just conceivably, in a book.

Further results and exercises

1. Analyse the data of Table 1.4 by the use of (a) unweighted and (b) weighted empirical logistic transforms. Compare the answers with the results of maximum likelihood fitting as given by Dyke and Patterson (1952) and Cox and Snell (1968).

[Sections 1.2, 2.8]

2. Examine the following fictitious data comparing two treatments; m is a number so large that sampling fluctuations can be ignored:

	Male		Female	
	Untreated	Treated	Untreated	Treated
Success	$4m$	$8m$	$2m$	$12m$
Total	$7m$	$13m$	$5m$	$27m$

Show that the treatment effect on a logistic scale is the same for males and females, but that if the data for the two sexes are pooled the success rate is the same for treated and untreated individuals. Express algebraically what is involved and comment on the implications for the analysis of binary data.

[Sections 1.2, 2.3, 2.8; Simpson, 1951; Lindley, 1964; Blyth, 1972]

3. The linear logistic model $\lambda = x\beta$ can be generalized in various ways, for example by

(a) adding non-linear terms in the explanatory variables, this possibly preserving the linear logistic structure;

(b) forming the projection pursuit model $\lambda = \sum f_k(x\beta_k)$ where the f_k are arbitrary non-linear functions and β_k represent directions on which the explanatory variables are to be projected;

(c) specializing (b) to $\lambda = \alpha + \sum f_s(x_s)$, where the arbitrary functions f_s are now each a function of a single component variable.
Discuss in general terms the scientific reasonableness and flexibility of these approaches. The formal fitting of (b) and (c) involves a combination of smoothing and local likelihood ideas.
[Sections 1.3, 2.6; Hastie and Tibshirani, 1987a, b; Tibshirani and Hastie, 1987]

4. It is required to approximate to the function $e^x/(1 + e^x)$, for $x \geqslant 0$, by (a) a normal distribution function, $\Phi(ax)$ and (b) by a function min $(bx, 1)$. Find the values of the constants a and b giving best fits in the following senses: (i) identity of tangents at $x = 0$; (ii) agreement of standard deviations of full distributions; (iii) Tchebychev (minimax) fit; (iv) least squares fit.

[Section 1.5]

5. In a binary stimulus–response situation the probability of success at stimulus x is $G(x)$, where $G(x)$ has the mathematical properties of a cumulative distribution function. Prove that by a suitable monotonic transformation of the stimulus scale, say to $z = z(x)$, the probability of success can be taken to be (a) $e^z/(1 + e^z)$, (b) $\Phi(z)$, the standardized normal integral. Show how the transformation for (a) can be estimated from data (i) for arbitrary functions, $G(x)$ and (ii) by assuming that it suffices to take a transformation in the family $z_\rho(x) = (x^\rho - 1)/\rho$ followed by a linear transformation.

[Section 1.5]

6. Prove that if R is the number of successes in n independent trials with probability of success θ, then for large n

$$\text{var}\left[\sin^{-1}\{\sqrt{(R/n)}\}\right] \sim 1/(4n),$$

the angle being measured in radians. Numerical studies show that

$$\sin^{-1}\sqrt{\{R/(n+1)\}} + \sin^{-1}\sqrt{\{(R+1)/(n+1)\}}$$

has a variance within $\pm 6\%$ of $1/(n + \frac{1}{2})$, when $n\theta \geqslant 1$. Examine this analytically by a Taylor expansion about $R = n\theta$. Discuss the circumstances under which a linear model associated with this transformation yields substantial computational advantages. What are its disadvantages, and what are the circumstances under which the more nearly constant variance associated with the second form is a practical advantage?

[Section 1.5; Freeman and Tukey, 1950]

7. Observations are made on k binary populations classified in a two-way arrangement with k rows and l columns. There are n_{ij} observations in the (i, j)th cell, the corresponding probability being θ_{ij} with logistic transform λ_{ij}. Obtain sufficient statistics under the additive model $\lambda_{ij} = \mu + \rho_i + \gamma_j$ and show how absence of interactions can be tested. If the n_{ij}'s are equal, or more generally of the form $n_{i.}n_{.j}$, discuss the advantages and limitations of an analysis in terms of an additive model on $\sin^{-1}\sqrt{\theta_{ij}}$.   [Sections 1.5, 2.8, 2.10]

8. Discuss the relation between the analysis of binary data and that of survival data, in particular the consequences of modifying the likelihood based theory of the Kaplan–Meier product-limit estimate by assuming that the conditional probabilities of failure depend at each stage logistically on explanatory variables.

[Section 1.6; Cox and Oakes, 1984; Efron, 1988]

9. Suppose that in a binary stimulus–response situation, the probability of success at stimulus x rises monotonically from a lower asymptote ξ, where $\xi \geqslant 0$, to an upper asymptote η, where $\eta \leqslant 1$. Show that a curve of this type is obtained when the probability of success at stimulus x is

$$\frac{\xi + \eta e^{\alpha + \beta x}}{1 + e^{\alpha + \beta x}},$$

where $\beta > 0$. Show how, from suitable data, the parameters ξ, η, α and β can be estimated. How can initial values be obtained for an iterative procedure? What special problems would arise in applying maximum likelihood theory to test the null hypothesis $\xi = 0$?

[Sections 1.3, 1.6, 3.4; Chernoff, 1954]

10. Show that the continuous logistic density $e^x/(1 + e^x)^2$ has moment generating function $s\pi/\sin(s\pi)$ and hence that the odd order cumulants are zero and that the $2r$th cumulant is

$$\frac{(-1)^{r+1} 2^{2r-1} B_{2r} \pi^{2r}}{r},$$

where B_{2r} is a Bernoulli number. In particular, show that the standard deviation is $\pi/\sqrt{3}$ and that the coefficient of kurtosis is $6/5$.

[Section 1.3; Abramowitz and Stegun, 1965, pp. 75, 810]

11. Interpret binary response curves in terms of an underlying continuous distribution in a way alternative to that involving a tolerance distribution. For this, suppose that there is an underlying continuous random variable W and that the binary response $Y = 1$ if and only if $W > 0$. Suppose further that corresponding to stimulus level x, the random variable W has a logistic distribution of mean $\alpha + \beta x$ and unit scale parameter. Generalize this to distributions other than the logistic, and to the case of the general linear model.

[Section 1.3]

12. Obtain estimates of a local linear regression in which the probability of success for the ith trial is $\gamma_0 + \gamma_i x_i$ and in which ordinary unweighted least squares formulae are applied to the $(0, 1)$ observations. Develop maximum likelihood and least squares formulae for this model (a) not subject to and (b) subject to the constraint that all fitted values are in $[0, 1]$.

[Sections 2.1, 2.2]

13. Prove that for the linear logistic model for the 2×2 contingency table, the maximum likelihood estimates and their standard errors are those given by analysis of the crude empirical logistic transforms Z and their associated variances V as defined in Section 2.1. Show that an analogous result holds for any linear logistic model saturated with parameters.

[Sections 2.1, 2.3]

14. In n independent trials with probability of success θ, the observed number of successes is r. If the prior density of θ is proportional to $\theta^a(1 - \theta)^b$, show that the posterior density is proportional to $\theta^{a+r}(1 - \theta)^{b+n-r}$. Hence show that the posterior density of

$$\left(\frac{n - r + b + 1}{r + a + 1} \right)\left(\frac{\theta}{1 - \theta} \right)$$

has the variance-ratio distribution with $(2r + 2a + 2,\ 2n - 2r + 2b + 2)$ degrees of freedom. Hence prove that the posterior distribution of $\log\{\theta/(1 - \theta)\}$ is approximately normal with mean and variance

$$\log\left(\frac{r+a+\frac{1}{2}}{n-r+b+\frac{1}{2}}\right) \text{ and } \frac{1}{r+a+1} + \frac{1}{n-r+b+1}.$$

Show that the methods of analysis using least squares and the empirical logistic transform therefore have a Bayesian interpretation involving independent prior distributions with small a and b.

[Sections 2.1, 3.3; Lindley, 1964]

15. Let R be the number of successes in n independent trials with constant probability of success θ. Prove that

$$E\left\{\psi\left(\frac{R+a}{n+b}\right)\right\} = \psi(\theta) + \frac{a-b\theta}{n}\psi'(\theta) + \frac{\theta(1-\theta)}{2n}\psi''(\theta) + o\left(\frac{1}{n}\right).$$

Examine the special cases $\psi(\theta) = \log\{\theta/(1-\theta)\}$, $\log\theta$ and $\sin^{-1}\sqrt{\theta}$ and hence obtain an approximately unbiased estimate of $\psi(\theta)$. Discuss critically the relevance of the requirement of unbiasedness.

[Section 2.1]

16. In an experiment at three stimulus levels, there are R_j successes in n_j trials at stimulus level x_j $(j = 1, 2, 3)$. Show that if the binary response curve is logistic, the statistic

$$(x_2 - x_3)\log\left(\frac{R_1 + \frac{1}{2}}{n_1 - R_1 + \frac{1}{2}}\right) + (x_3 - x_1)\log\left(\frac{R_2 + \frac{1}{2}}{n_2 - R_2 + \frac{1}{2}}\right)$$

$$+ (x_1 - x_2)\log\left(\frac{R_3 + \frac{1}{2}}{n_3 - R_3 + \frac{1}{2}}\right)$$

has asymptotically zero expectation and a variance estimated by

$$\sum_{i>j\neq k} (x_i - x_j)^2 \frac{(n_k + 1)(n_k + 2)}{n_k(R_k + 1)(n_k - R_k + 1)}.$$

Does this lead to an asymptotically unique test of the adequacy of the logistic model? Give the corresponding test statistic for an integrated normal response curve.

[Section 2.1; Chambers and Cox, 1967]

17. Examine an 'exact' procedure for the comparison of two groups adjusting for a concomitant variable z, assuming logistic transforms $\alpha + \beta z$ and $\alpha + \beta z + \Delta$ in the two groups.

[Sections 2.1, 2.2]

18. In n mutually independent binary trials the probability of success is θ_0 for trials $1, \ldots, v$ and is θ_1 for trials $v+1, \ldots, n$, where v is unknown, but θ_0 and θ_1 are known. Find the likelihood and show that the maximum likelihood estimate of v is obtained by scoring $\log\{(1 - \theta_0)/(1 - \theta_1)\}$ for a failure and $\log(\theta_0/\theta_1)$ for a success. Then $\hat{v}$ is such that the cumulative score from observations $1, \ldots, m$, is a maximum at $m = \hat{v}$. Set out the corresponding procedure when θ_0 and θ_1 are unknown.

[Section 2.1; Page, 1955]

19. Analyse the 2×2 contingency table of Table 1.2 by the empirical logistic transform of Section 2.1.6. That is, test the hypothesis $\Delta = 0$ via the difference of z's divided by its estimated standard error and obtain confidence intervals via the pivot $(z_1 - z_0 - \Delta)/\sqrt{(v_1 + v_0)}$. Show that the significance level corresponds closely to that from the use of chi-squared without a continuity correction.

[Section 2.1]

20. Investigate the application of the ideas of D-optimality to the design of experiments for estimating the parameters in a simple linear logistic model, it being assumed that rough initial estimates are available for use in fixing the design points. Assuming that the optimal design is a two-point design symmetrically placed around the 50% point determine the specific points to be used.

[Section 2.1; Minkin, 1987]

21. Develop a Wald likelihood ratio sequential test for testing the hypothesis that the probability of success is a known constant, against the alternative that there is a linear trend of specified amount on a logistic scale, starting from the known initial value. Obtain also a test for the corresponding problem when the constant probability is an unknown nuisance parameter.

[Section 2.2; Wald, 1947; Cox, 1963]

22. In the problem of comparing probabilities with observed successes and failures, suppose that a concomitant measurement is available for each individual. How can this be used? Given two sequences $p_1, \ldots, p_n; p'_1, \ldots, p'_n$ of probabilities, both of which fit the data adequately, suggest ways of deciding which is the more informative sequence.

[Section 2.2; Cox, 1958b]

23. Suppose that in an analysis for trend in the probability of success, the null hypothesis is that the probability is a known constant. Formulate this as the problem of testing $\beta = 0$ when the logistic transform of the probability of success on the ith trial is $\alpha_0 + \beta(i - 1)$, and hence obtain the form of the exact test and the mean and variance of the test statistic under the null hypothesis. Compare the test with that appropriate when the initial probability is unknown. Discuss the analysis when there are several sequences, and departures from the null hypothesis may occur both in initial value and in trend, indicating a scheme for graphical and formal analysis.

[Section 2.2; Cox, 1958a]

24. Prove that for the serial regression problem with n trials and r successes the null probability generating function of the test statistic is

$$\frac{1}{\binom{n}{r}} \prod_{i=1}^{r} \left(\frac{\zeta^i - \zeta^{n+1}}{1 - \zeta^i} \right).$$

[Section 2.2; Haldane and Smith, 1948]

25. Justify the following procedure for fitting a logistic stimulus-response curve to equally spaced stimulus levels. Plot the crude empirical logistic transforms against stimulus level. If the result is not exactly linear, adjust the data as follows: for a suitable stimulus level, subtract $2a$ successes from one stimulus level and add a successes at both neighbouring stimulus levels, where a is a suitable real number, not necessarily an integer. Replot and if necessary repeat the process of adjustment until the plotted points lie effectively on a straight line, the position and slope of the line giving the maximum likelihood estimates.

[Section 2.2; Hodges, 1958]

26. Prove that the maximum likelihood equations for the stimulus-response model $\theta_i = e^{\alpha + \beta x_i}/(1 + e^{\alpha + \beta x_i})$ can be written, in the usual notation,

$$\sum R_i = \sum n_i \hat{\theta}_i, \qquad \sum R_i x_i = \sum n_i x_i \hat{\theta}_i,$$

where $\hat{\theta}_i$ is obtained by substituting $\hat{\alpha}$ and $\hat{\beta}$ into the expression for θ_i. Examine the special case where the x_i's are equally and narrowly spaced at intervals h, over a range from very small θ_i to θ_i very near 1, and where all the n_i's are equal. Show, by replacing sums by integrals, that if $\mu = -\alpha/\beta$ is the 50% point, then approximately

$$\hat{\mu} = x_{\max} + \tfrac{1}{2}h - h\sum R_i/n,$$

$$\hat{\mu}^2 + \frac{\pi^2}{3\hat{\beta}^2} = (x_{\max} + \tfrac{1}{2}h)^2 - \tfrac{1}{12}h^2 - 2h\sum R_i x_i/n.$$

Here $\hat{\mu}$ is called the Spearman–Kärber estimate.

[Section 2.2; Anscombe, 1956]

27. It is required to determine sequentially the value of the explanatory variable x at which the probability of success is equal to a preassigned value p, say. Suppose this is done by assuming a linear logistic relation, estimating the parameters by maximum likelihood on the basis of the current data and placing the next observation at the current estimate of the p point. Relate this to the Robbins–Monro procedure for continuous responses and show desirable properties of consistency under a general model and efficiency under the assumed logistic model.

[Section 2.2; Wu, 1985]

28. Individuals are allocated at random to two treatments, a preassigned number n_0 receiving treatment 0 and the remaining $n_1 = n - n_0$ receiving treatment 1. On each individual a binary response is then observed. The null hypothesis is that the response observed on any individual is unaffected by the treatment applied to that and other individuals; thus, in particular, the total number r of successes does not depend on the treatment allocation. Show that a test of the null hypothesis based on the randomization distribution of the number of successes in the first treatment is formally identical with Fisher's exact test for the 2×2 contingency table. Show that the other 'exact' tests of Section 2.3 also can be regarded as randomization tests.

[Sections 2.3, 4.5; Barnard, 1947]

29. Consider a generalization of the matched pair situation in which there are t treatments and the observations are grouped into matched sets of t, one individual in each set receiving each treatment. Set up a logistic model and show that the total numbers of successes per treatment and per set, form the sufficient statistics. An overall test of the equivalence of treatments may be based on the corrected sum of squares of the treatment 'totals'; obtain the mean and variance of the

test statistic under the null hypothesis and suggest a chi-squared approximation.

[Section 2.2; Cochran, 1950]

30. Prove that if X has a binomial distribution of index n and parameter θ, then $E\{X(n - X)\} = n(n - 1)\theta(1 - \theta)$.

To compare two treatments $n = n_0 + n_1$, individuals are tested n_0 for treatment 0 and n_1 for treatment 1, and a binary response observed. Show that conditionally on the total number t of successes the Fisher information about Δ, the logistic difference, evaluated at $\Delta = 0$, is

$$\frac{n_0 n_1 (n_0 + n_1 - t)t}{(n_0 + n_1)^2(n_0 + n_1 - 1)}.$$

Note that near $\Delta = 0$, t is the observed value of a binomial random variable of index $n_0 + n_1$ and parameter θ, say; hence show that the expected Fisher information is

$$n_0 n_1 \theta(1 - \theta)/(n_0 + n_1).$$

Show further that if the $n_0 + n_1$ individuals are randomized between the two treatments the expected Fisher information (i.e. before the outcome of the randomization is observed) is $\frac{1}{4}(n - 1)\theta(1 - \theta)$, whereas if exact balance is enforced the corresponding value is $\frac{1}{4}n\theta(1 - \theta)$.

[Section 2.3; Cox, 1988]

31. For the 2×2 table with probabilities of success ϕ_s and R_s, successes observed out of n_s trials ($s = 0, 1$) show that the sample cross-product ratio has the property that

$$\{E(R_0)E(n_1 - R_1)\}/\{E(n_0 - R_0)E(R_1)\} = \psi$$

where $\psi = \phi_0(1 - \phi_1)/\{(1 - \phi_0)\phi_1\}$ if it is assumed that the data are from two independent binomal populations with n_0 and n_1 fixed, whereas if all four marginal totals are assumed fixed then

$$\{E(R_0)E(n_1 - R_1) + \text{var}(R_0)\}/\{E(n_0 - R_0)E(R_1) + \text{var}(R_0)\} = \psi$$

hence indicating that the conditional maximum likelihood estimate of the odds ratio is weighted towards unity.

[Section 2.3; Mantel and Hankey, 1975]

32. For matched pair data, assuming the logistic model (2.57) and corresponding binomial probability (2.58), show that the conditional maximum likelihood estimate of the logistic difference is

$$\hat{\Delta}_c = \log(R^{01}/R^{10}),$$

where R^{01} and R^{10} are the numbers of pairs with mixed responses, whereas the unconditional maximum likelihood estimate is

$$\hat{\Delta} = 2\log(R^{01}/R^{10}).$$

[Sections 2.4, 2.5.4; Andersen, 1973a, p. 69; Breslow and Day, 1980, p. 249]

33. Develop a random effects analysis of the paired comparison situation in which for the sth pair, the probabilities of success are the normal integrals with arguments

$$\mu + U_s \quad \text{and} \quad \mu + \Delta + U_s,$$

where U_s is a random variable, characteristic of the sth pair, normally distributed with zero mean and variance σ_u^2, different U_s's being mutually independent. Prove that the probabilities of the four types of response $(0,0), (0,1), (1,0), (1,1)$ are given by integrals of the bivariate normal distribution and hence develop a procedure for estimating Δ, together with a standard error of estimation.

[Sections 2.4, 3.4]

34. In a matched pair comparison, a concomitant variable u_s is associated with both individuals in the sth pair. Suppose that in the model (2.57) in which a parameter α_s is associated with the sth pair, it is assumed that $\alpha_s = \alpha + \beta u_s$. Discuss the problems of exact and approximate inference about the difference between treatments. How is the analysis modified if the concomitant variable is different for the two individuals in a pair?

[Section 2.4]

35. In two 2×2 contingency tables comparing the same two treatments, 0 and 1, the logistic differences between 0 and 1 are Δ_0 and Δ_1. In the sth table the observed numbers of successes are r_{s0} and r_{s1} and the numbers of trials n_{s0} and n_{s1} $(s = 1, 2)$. Show that for

inferences about $\Delta = \Delta_1 - \Delta_0$ the distribution for use is

$$p_{R_{11}}(t;\Delta) = \frac{c(n,r,t)\Delta^t}{\sum_u c(n,r,u)\Delta^u},$$

where

$$c(n,r,t) = \binom{n_{00}}{r_{00} - r_{11} + t}\binom{n_{01}}{r_{00} + r_{01} - t}\binom{n_{10}}{r_{10} + r_{11} - t}\binom{n_{11}}{t}.$$

[Sections 2.5, 2.8; Bartlett, 1935]

36. Suppose that a large number of 2×2 tables are available comparing the same two treatments. A plot of the estimated logistic difference, Δ, between treatments against the mean logistic transform μ for the two treatments, shows a systematic relation, in which, approximately, $\Delta = g(\mu)$. Show that if the differences Δ are small an analysis of the differences of $h[\log\{\theta/(1-\theta)\}]$, where $h'(x)g(x) = \text{const}$, leads approximately to constant treatment differences.

[Section 2.5]

37. Discuss the disadvantages and advantages, if any, of the following methods of analysing k 2×2 tables, as compared with the methods of Section 2.5; (i) computation of a significance level P_s from the sth table and their combination into the statistic $-2\sum \log P_s$, to be tested approximately as chi-squared with $2k$ degrees of freedom; (ii) computation of a standard chi-squared statistic from each table, not corrected for continuity, followed by the testing of the sum as chi-squared with k degrees of freedom; (iii) computation of a signed chi statistic from each table, followed by a test of their sum as normally distributed, with zero mean and variance k.

[Section 2.5; Yates, 1955]

38. Discuss critically the advantages and disadvantages of the squared multiple correlation coefficient, R^2, for summarizing for normal-theory linear regression (a) the adequacy of two different models fitted to the same data, (b) the adequacy of the same model fitted to two independent sets of data. Develop a similar criterion for more general models, including linear logistic models fitted to binary data. For example, suppose that $\hat{L}_b$ and $\hat{L}_f$ are the maximized

likelihoods of a baseline model and a model under study; the baseline model might, for example, correspond to assuming independent and identically distributed observations. Note that $P = (\hat{L}_f/\hat{L}_b)^{1/n}$ is for n observations, the geometric mean improvement per observation produced by fitting the more elaborate model and that for the normal-theory linear model $R^2 = 1 - 1/P$.

[Section 2.6]

39. If $\hat{p} = \exp(\hat{\lambda})/\{1 + \exp(\hat{\lambda})\}$ and $\hat{\gamma}$ denote the maximum likelihood estimates obtained by fitting the augmented logistic model $\lambda = X\beta + z\gamma$ to observations $y_i = 0, 1$ for $i = 1, \ldots, n$ show that

$$z\hat{\gamma} + (y - \hat{p})/\{\hat{p}(1 - \hat{p})\}$$

defines a partial residual (for use in checking the form of the dependence on the explanatory variable z).

Hint: if the logistic model $\lambda = X\beta$ is fitted using the Newton–Raphson iterative method, the $(t + 1)$th iteration is given by

$$\hat{\beta}_{(t+1)} = \hat{\beta}_{(t)} + (X^T V X)^{-1} X^T (y - \hat{p})$$
$$= (X^T V X)^{-1} X^T V \{X\hat{\beta}_{(t)} + V^{-1}(y - \hat{p})\},$$

where $V = \mathrm{diag}\, \hat{p}_i(1 - \hat{p}_i)$. By analogy with standard least squares equations $X\hat{\beta}_{(t)} + V^{-1}(y - \hat{p})$ can be regarded as the vector of observations and $V^{-1}(y - \hat{p})$ as the residual vector. Similar reasoning for the augmented model leads to the required answer.

[Section 2.7; Landwehr, Pregibon and Shoemaker, 1984]

40. Let R have a binomial distribution corresponding to n trials with probability of success θ. Obtain by expansion the asymptotic skewness of $\phi(Y/n)$, for an arbitrary function $\phi(u)$. Prove that this vanishes if

$$\phi(u) = \int_0^u t^{-1/2}(1 - t)^{-1/2} dt.$$

Hence show that

$$[\phi(Y/n) - \phi\{\theta - \tfrac{1}{6}(1 - 2\theta)/n\}]/\{\theta^{1/2}(1 - \theta)^{1/2}/\sqrt{n}\}$$

may be expected to have a distribution close to the standardized normal distribution. Check this numerically for $n = 5$ and 10; $\theta = 0.1$.

[Section 2.7; Blom, 1954; Cox and Snell, 1968]

41. Let R_i have a binomial distribution with n_i trials and with probability of success $\theta_i(\beta)$, where β is an unknown parameter. Let $\hat{\beta}$ be the maximum likelihood estimate of β and write $\hat{\theta}_i = \theta_i(\hat{\beta})$. Show that residuals could be defined, for any function $\psi(u)$, by

$$\frac{\psi(R_i/n_i) - \psi(\hat{\theta}_i)}{|\psi'(\hat{\theta}_i)|\{\hat{\theta}_i(1 - \hat{\theta}_i)/n_i\}^{1/2}}.$$

Discuss the relative merits of the choices $\psi(u) = u$, $\psi(u) = \phi(u)$, as given in Exercise 40, and, for linear logistic models, $\psi(u) = \log\{u/(1 - u)\}$.

[Section 2.7; Cox and Snell, 1968]

42. Obtain, for the situation of Exercise 7, a test broadly analogous to Tukey's degree of freedom for non-additivity, by considering the statistic

$$\sum R_{is} \frac{R_{i\cdot} R_{\cdot s}}{n_{i\cdot} n_{\cdot s}},$$

where R_{is} is the number of successes in the (i, s)th cell. Show how to obtain, conditionally on $R_{i\cdot} = r_{i\cdot}$, $R_{\cdot s} = r_{\cdot s}$, the mean and variance of the test statistic.

[Section 2.8; Tukey, 1949]

43. Set up a linear logistic model for the matched pair problem in which there is an additional parameter representing an effect of the order in which the two treatments are applied. Obtain a significance test for the treatment effect.

[Section 2.9; Gart, 1969]

44. Consider a binary time series in which the probability of success at any trial depends on the outcomes of the previous q trials. Show that for a general q-dependent Markov chain, 2^q parameters have to be estimated, whereas if the logistic analogue of a qth order autoregressive process is taken, only $q + 1$ parameters are involved. Develop procedures of estimation and testing associated with these two models.

[Section 2.11]

45. Suppose that in a sequence of trials $\lambda_s = \alpha + \beta'\cos(s\omega) + \beta''\sin(s\omega)$ for known ω. Show that the sufficient statistics are $\sum Y_s$,

$\sum Y_s \cos(s\omega)$ and $\sum Y_s \sin(s\omega)$. If the phase of the relationship is irrelevant, these can be reduced to $\sum Y_s$ and

$$K(\omega) = \{\sum Y_s \cos(s\omega)\}^2 + \{\sum Y_s \sin(s\omega)\}^2,$$

which is proportional to the periodogram ordinate as commonly defined. Conditionally on the observed value of $\sum Y_s$, find the mean and variance of $K(\omega)$ when $\beta' = \beta'' = 0$ and $\omega = 2\pi r/n$, for integral r.

[Section 2.11]

46. In a series of n binary trials the probability of success at any trial, given that there have been v successes in the previous trials is $e^{\alpha + \beta v}/(1 + e^{\alpha + \beta v})$, i.e. each success increases the logistic transform by β. Let s_0 be the serial number of the first success, $s_0 + s_1$ that of the second success, and $s_0 + \cdots + s_{r-1}$ that of the rth and final success; also write $s_r = n - s_0 - \cdots - s_{r-1}$. Prove that the likelihood is

$$\frac{\exp\{r\alpha + \tfrac{1}{2}r(r-1)\beta\}}{\prod\limits_{i=0}^{r}(1 + e^{\alpha + i\beta})^{s_i}}.$$

Hence show that there is no simple sufficient statistic. Show further that if β is large or small, the likelihood involves the s_i's through $T = \sum i s_i$. Prove that when $\beta = 0$ the mean and variance of T are

$$\tfrac{1}{2}(r'-1)n \quad \text{and} \quad \frac{r'(r'+1)}{12}\sum(s_i - \bar{s})^2,$$

where $r' = r + 1$ if the series ends with a failure and $r' = r$ otherwise.

[Section 2.11; Cox, 1958a]

47. Find the likelihood for the generalization of Exercise 46 in which each success increases the logistic transform by β and each failure decreases it by γ.

[Section 2.11]

48. A parameter-driven process is defined (approximately) as follows: the unobserved probabilities $\{\theta_t\}$ are governed by a stationary linear first-order autoregressive process, i.e.

$$\theta_{t+1} = \mu + \rho(\theta_t - \mu) + \eta_{t+1},$$

where $\{\eta_t\}$ are independently identically distributed with zero mean and variance σ^2, $|\rho| < 1$, and where μ, ρ and σ^2 are such that θ_t takes high probability values in $(0, 1)$. Given $\{\theta_t\}$, the observations $\{Y_t\}$ are independent binary random variables with $E(Y_t) = \theta_t$.

Prove that

$$E(Y_t) = \mu, \quad E(Y_t Y_{t+r}) = \text{prob}(Y_t = Y_{t+r} = 1) = \mu^2 + \rho^r \sigma^2/(1 - \rho^2).$$

Hence suggest how μ, ρ, σ^2 might be estimated. How would you investigate the efficiency of these estimates?

To study the adequacy of the model, and in particular to distinguish it from a one- or two-dependent Markov chain, we may use the second-order transition probabilities. Calculate

$$\text{prob}(Y_{t+1} = 1 \mid Y_t = Y_{t-1} = 1),$$
$$\text{prob}(Y_{t+1} = 1 \mid Y_t = 0, \ Y_{t-1} = 1),$$
$$\text{prob}(Y_{t+1} = 1 \mid Y_t = 1, \ Y_{t-1} = 0),$$
$$\text{prob}(Y_{t+1} = 1 \mid Y_t = Y_{t-1} = 0) \quad \text{and}$$

suggest how these can be used to test the adequacy of the model.

[Section 2.11]

49. Show that the beta-binomial distribution for suitable choice of parameters is consistent with a limited range of forms of underdispersion.

[Section 3.2; Prentice, 1986]

50. In n independent trials the probabilities of success are $\theta_1, \ldots, \theta_n$ and T is the total number of successes. Prove that if and only if $\sum \theta_i(1 - \theta_i)$ diverges as $n \to \infty$, the distribution of T is asymptotically normal. Calculate γ_1 and γ_2 for the distribution of T and show how to use these to improve the normal approximation. Give an algorithm suitable for computing 'exact' tail areas from the probability generating function of T.

[Section 3.2]

51. In a stimulus–response situation, there are target stimulus levels $x_1, \ldots, x_n$, there being one observation at each level. The stimuli actually applied are unknown but are independently normally distributed around the targets with variance τ^2. Obtain an expression for the probability of success at target stimulus x, assuming that the logistic regression on actual dose ξ is $\alpha + \beta\xi$. Show that if $\beta\tau$ is small,

the logistic transform at target stimulus x is approximately

$$\alpha + \beta x + \tfrac{1}{2}\tau^2 \beta^2 \frac{1 - e^{\alpha + \beta x}}{1 + e^{\alpha + \beta x}}.$$

Discuss the qualitative change in the shape of the response curve. Write down an approximate likelihood and outline an iterative procedure for estimating α and β, regarding τ^2 as known.

[Section 3.4]

52. Show that the situation of the previous exercise is more easily handled when the response relation is of the integrated normal type, when a probability of success at actual level ξ of $\Phi\{(\xi - \mu)/\sigma\}$ is converted into a probability $\Phi\{(x - \mu)/\sigma'\}$ at target level x, where $\sigma'^2 = \sigma^2 + \tau^2$. Prove this result analytically and by an argument based on tolerances.

[Section 3.4]

53. Contrast the situation of the previous two exercises with that in which there is a random error in measuring the stimulus applied, so that each measured stimulus x_i is normally distributed with variance τ^2 around the corresponding actual stimulus ξ_i, which is unknown. Discuss the difficulties of estimating an assumed logistic regression on ξ, even when τ^2 is known.

[Section 3.4]

54. Consider independent observations on a bivariate binary response (Y_1, Y_2), i.e. a 2×2 contingency table with both rows and columns corresponding to random variables, and let $\theta_{is} = \text{prob}(Y_1 = i, Y_2 = s)$, with $\theta_{i.}$ and $\theta_{.s}$ referring to the marginal distributions. Suppose that a measure of association of (Y_1, Y_2) is a function (a) of the pair of conditional probabilities $\theta_{00}/\theta_{0.}$, $\theta_{10}/\theta_{1.}$, and (b) of the pair $\theta_{00}/\theta_{.0}$, $\theta_{01}/\theta_{.1}$. Show that the measure is a function of $(\theta_{00}\theta_{11})/(\theta_{01}\theta_{10})$, i.e. of the logistic difference.

[Section 4.3; Edwards, 1963]

55. A non-null randomization theory of the 2×2 contingency table can be approached by classifying the n individuals into four (unobservable) groups: n_{00} failing under both treatments; n_{10} giving

success under T_0 and failure under T_1; n_{01} giving failure under T_0 and success under T_1; n_{11} giving success under both treatments. A very strong null hypothesis is that $n_{01} = n_{10} = 0$, i.e. every individual response is unaffected by treatment. The overall difference in the proportion of successes is

$$(n_{01} + n_{11})/n - (n_{10} + n_{11})/n = (n_{01} - n_{10})/n = \delta,$$

say. Suppose that of the n individuals n_0 are randomly allocated to T_0 and n_1 to T_1, $n_0 + n_1 = n$. Then if x_{is} is the number of individuals of the type is that are allocated to T_0, show that the x_{is} have a multivariate hypergeometric distribution. Note that the resulting observations specify the values of

$$x_{00} + x_{01} = n_0, \quad x_{10} + x_{11} = r_0, \quad (n_{01} - x_{01}) + (n_{11} - x_{11}) = r_1;$$

obtain the likelihood of the observations in terms of δ and two other parameters. There remains the difficulty of extracting information about δ from this.

[Section 4.5; Copas, 1973]

56. Interpret the Bradley–Terry paired comparison model in terms of an underlying continuous logistic distribution of response; assume that the response when P_i is compared with P_s is a logistic variate of unit dispersion and mean $\rho_i - \rho_s$ and that P_i is preferred to P_s if and only if this variate is positive. Show that in an analogous model using normal distributions, the probability that P_i is preferred to P_s is $\Phi(\xi_i - \xi_s)$, called the Thurstone model. Prove that in the Bradley–Terry model, but not in the Thurstone model, there are simple sufficient statistics for the unknown parameters.

[Section 5.2; David, 1988, Chapter 4]

57. For three categories of response discuss the relative merits of the models in Sections 5.3 and 5.4 for regression on a single explanatory variable, x.

[Sections 5.3, 5.4]

58. Derive an optimum discriminator between two populations of mixed binary and multivariate normal responses, setting up plausible distributional forms.

[Section 5.6]

References

Abdelbasit, K.M. and Plackett, R.L. (1981) Experimental design for categorized data. *Int. Statist. Rev.*, **49**, 111–26.

Abramowitz, M. and Stegun, I.A. (1965) *Handbook of Mathematical Functions*, Dover, New York.

Agresti, A. (1984) *Analysis of Ordinal Categorical Data*, Wiley, New York.

Agresti, A. (1988) A model for agreement between ratings on an ordinal scale. *Biometrics*, **44**, 539–48.

Aitchison, J. and Silvey, S.D. (1957) The generalization of probit analysis to the case of multiple responses. *Biometrika*, **44**, 131–40.

Altham, P.M.E. (1969) Exact Bayesian analysis of a 2 × 2 contingency table, and Fisher's 'exact' significance test. *J.R. Statist. Soc.*, **B 31**, 261–9.

Amemiya, T. (1985) *Advanced Econometrics*, Blackwell, Oxford.

Andersen, E.B. (1973a) Conditional inference and models for measuring. Mentalhygiejnisk, Copenhagen.

Andersen, E.B. (1973b) A goodness-of-fit test for the Rasch model. *Psychometrika*, **38**, 1223–140.

Andersen, E.B. (1980) *Discrete Statistical Models with Social Science Applications*. North-Holland, Amsterdam.

Anderson, J.A. (1984) Regression and ordered categorical variables (with discussion). *J.R. Statist. Soc.*, **B 46**, 1–30.

Anderson, J.A. and Philips, P.R. (1981) Regression, discrimination and measurement models for ordered categorical variables. *Appl. Statist.*, **30**, 22–31.

Anderson, S., Auquier, A., Hauck, W.W., Oakes, D., Vandaele, W. and Weisberg, H.I. (1980) *Statistical Methods for Comparative Studies*, Wiley, New York.

Andrews, D.F. and Herzberg, A.M. (1985) *Data: A Collection of Problems from Many Fields for the Student and Research Worker*, Springer-Verlag, New York.

Anscombe, F.J. (1956) On estimating binomial response relations. *Biometrika*, **43**, 461–4.

Aranda-Ordaz, F.J. (1981) On two families of transformations to additivity for binary response data. *Biometrika*, **68**, 357–63.

Armitage, P. (1955) Tests for linear trends in proportions and frequencies. *Biometrics*, **11**, 375–86.

Azzalini, A. (1983) Maximum likelihood estimation of order m for stationary stochastic processes. *Biometrika*, **70**, 381–7.

Barnard, G.A. (1947) Significance tests for 2×2 tables. *Biometrika*, **34**, 123–38.

Barndorff-Nielsen, O. and Cox, D.R. (1979) Edgeworth and saddle-point approximations with statistical applications (with discussion). *J.R. Statist. Soc.*, **B 41**, 279–312.

Bartlett, M.S. (1935) Contingency table interactions. *J.R. Statist. Soc. Suppl.*, **2**, 248–52.

Bartlett, M.S. (1937) Some examples of statistical methods of research in agriculture and applied biology (with discussion). *Suppl. J.R. Statist. Soc.*, **4**, 137–70.

Basawa, I.V. and Scott, D.J. (1983) Asymptotic optimal inference for non-ergodic models. *Lecture Notes in Statistics*, **17**, Springer-Verlag, New York.

Begg, C.B. and Gray, R. (1984) Calculation of polychotomous logistic regression parameters using individualized regressions. *Biometrika*, **71**, 11–18.

Berkson, J. (1944) Application of the logistic function to bio-assay. *J. Amer. Statist. Assoc.*, **39**, 357–65.

Berkson, J. (1951) Why I prefer logits to probits. *Biometrics*, **7**, 327–39.

Berkson, J. (1953) A statistically precise and relatively simple method of estimating the bio-assay with quantal response, based on the logistic function. *J. Amer. Statist. Assoc.*, **48**, 565–99.

Berkson, J. (1955a) Maximum likelihood and minimum χ^2 estimates of the logistic function. *J. Amer. Statist. Assoc.*, **50**, 130–62.

Berkson, J. (1955b) Estimation by least squares and by maximum likelihood. *Proc. 3rd Berkeley Symp.*, **1**, 1–11.

Berkson, J. (1957) Tables for the maximum likelihood estimate of the logistic function. *Biometrics*, **13**, 28–34.

Berkson, J. (1960) Nomograms for fitting the logistic function by maximum likelihood. *Biometrika*, **47**, 121–41.

Berkson, J. (1968) Application of minimum logit χ^2 to a problem of

Grizzle with a notation on the problem of no interaction. *Biometrics*, **24**, 75–95.

Bhapkar, V.P. (1961) Some tests for categorical data. *Ann. Math. Statist.*, **32**, 72–83.

Bhapkar, V.P. (1968) On the analysis of contingency tables with a quantitative response. *Biometrics*, **24**, 329–38.

Bhapkar, V.P. and Koch, G.G. (1968) On the hypotheses of no interaction in contingency tables. *Biometrics*, **24**, 567–94.

Billingsley, P. (1961) Statistical methods in Markov chains. *Ann. Math. Statist.*, **32**, 12–40.

Bishop, Y.M.M., Fienberg, S.E. and Holland, P.W. (1975) *Discrete Multivariate Analysis*, MIT Press, Cambridge, Mass.

Blom, G. (1954) Transformations of the binomial, negative binomial, Poisson and χ^2 distributions. *Biometrika*, **41**, 302–16.

Bloomfield, P. (1974) Transformations for multivariate binary data. *Biometrics*, **30**, 609–17.

Blyth, C.R. (1972) On Simpson's paradox and the sure thing principle, and some probability paradoxes in the choice from many random alternatives (with discussion). *J. Amer. Statist. Assoc.*, **67**, 364–81.

Bradley, R.A. (1976) Science, statistics and paired comparisons. *Biometrics*, **32**, 213–32.

Bradley, R.A. and El-Helbawy, A.T. (1976). Treatment contrasts in paired comparisons: basic procedures with application to factorials. *Biometrika*, **63**, 255–62.

Breslow, N.E. (1976) Regression analysis of the log odds ratio: a method for retrospective studies. *Biometrics*, **32**, 409–16.

Breslow, N.E. (1981) Odds ratio estimators when the data are sparse. *Biometrika*, **88**, 73–84.

Breslow, N.E. and Cologne, J. (1986) Methods of estimation in log odds ratio regression models. *Biometrics*, **42**, 949–54.

Breslow, N.E. and Day, N. (1980) *Statistical Methods in Cancer Research, 1: The Analysis of Case-control Studies*. IARC, Lyon.

Bross, I. (1954) Misclassification in 2 × 2 tables. *Biometrics*, **10**, 478–86.

Brown, B.W. (1980) Prediction analyses for binary data. In *Biostatistics Casebook* (eds. R.J. Miller, B. Efron, B.W. Brown, and L.E. Moses) Wiley, New York, pp. 3–18.

Carroll, R.J., Spiegelman, C.H., Lan, K.K., Bailey, K.T. and Abbott, R.D. (1984) On errors-in-variables for binary regression models. *Biometrika*, **71**, 19–25.

Chambers, E.A. and Cox, D.R. (1967) Discrimination between alternative binary response models. *Biometrika*, **54**, 573–8.

Chernoff, H. (1954) On the distribution of the likelihood ratio. *Ann. Math. Statist.*, **25**, 573–8.

Claringbold, P.J. Biggers, J.D. and Emmens, C.W. (1953) The angular transformation in quantal analysis. *Biometrics*, **9**, 467–84.

Cochran, W.G. (1950) The comparison of percentages in matched samples. *Biometrika*, **37**, 256–66.

Cook, R.D. (1986) Assessment of local influence (with discussion). *J.R. Statist. Soc.*, **B48**, 133–69.

Copas, J.B. (1973) Randomization models for the matched and unmatched 2 × 2 tables. *Biometrika*, **60**, 467–76.

Copas, J.B. (1988) Binary regression models for contaminated data (with discussion). *J.R. Statist. Soc.*, **B50**, 225–65.

Cornfield, J. (1956) A statistical problem arising from retrospective studies. *Proc. 3rd Berkeley Symp.*, **4**, 135–48.

Cornfield, J. (1962) Joint dependence of the risk of coronary heart disease on serum cholesterol and systolic blood pressure: a discriminant function analysis. *Fed. Proc.*, **21**, 58–61.

Cox, D.R. (1958a) The regression analysis of binary sequences (with discussion). *J.R. Statist. Soc.*, **B20**, 215–42.

Cox, D.R. (1958b) Two further applications of a model for binary regression. *Biometrika*, **45**, 562–5.

Cox, D.R. (1963) Large sample sequential tests of composite hypotheses. *Sankhyā*, **A25**, 5–12.

Cox, D.R. (1966a) A simple example of a comparison involving quantal data. *Biometrika*, **53**, 215–20.

Cox, D.R. (1966b) Some procedures connected with the logistic qualitative response curve. *Research Papers in Statistics: Essays in Honour of J. Neyman's 70th Birthday* (ed. F.N. David), Wiley, London, pp. 55–71.

Cox, D.R. (1972) The analysis of multivariate binary data. *Appl. Statist.*, **21**, 113–20.

Cox, D.R. (1975) Prediction intervals and empirical Bayes confidence intervals. In *Perspectives in Probability and Statistics.* (ed. J. Gani), Academic Press, London, pp. 47–55.

Cox, D.R. (1983) Some remarks on overdispersion. *Biometrika*, **70**, 269–74.

Cox, D.R. (1988) A note on design when response has an exponential family distribution, *Biometrika*, **75**, 161–4.

Cox, D.R. and Hinkley, D.V. (1974) *Theoretical Statistics*, Chapman and Hall, London.

Cox, D.R. and Lewis, P.A.W. (1966) *The Statistical Analysis of Series of Events*, Methuen, London.

Cox, D.R. and Oakes, D. (1984) *Analysis of Survival Data*, Chapman and Hall, London.

Cox, D.R. and Snell, E.J. (1968) A general definition of residuals (with discussion). *J.R. Statist. Soc.*, **B30**, 248–75.

Cox, D.R. and Snell, E.J. (1971). On test statistics calculated from residuals. *Biometrika*, **58**, 589–94.

Cox, D.R. and Snell, E.J. (1981) *Applied Statistics*, Chapman and Hall, London.

Crowder, M.J. (1978) Beta-binomial anova for proportions. *Appl. Statist.*, **27**, 34–7.

Daniel, C. (1959) Use of half-normal plots in interpreting factorial two-level exponents. *Technometrics*, **1**, 311–41.

Darroch, J.N. (1962) Interactions in multi-factor contingency tables. *J.R. Statist. Soc.*, **B24**, 251–63.

Darroch, J.N. (1974) Multiplicative and additive interaction in contingency tables. *Biometrika*, **61**, 207–14.

David, H.A. (1988) *The Method of Paired Comparisons*, 2nd edn, Griffin, London.

Davidson, R.R. and Beaver, R.J. (1977) Extending the Bradley–Terry model to incorporate within-pair order effects. *Biometrics*, **33**, 693–702.

Davidson, R.R. and Farquhar, P.H. (1976) A bibliography on the method of paired comparisons. *Biometrics*, **32**, 241–52.

Davison, A.C. (1988) Approximate conditional inference in generalized linear models. *J.R. Statist. Soc.*, **B50**, to appear.

Donner, A. and Hauck, W.W. (1986) The large-sample relative efficiency of the Mantel–Haenszel estimator in the fixed-strata case. *Biometrics*, **42**, 537–45.

Donner, A. and Hauck, W.W. (1988) Estimation of a common odds ratio in case-control studies of familiar aggregation. *Biometrics*, **44**, 369–78.

Dorn, H.F. (1954) The relationship of cancer of the lung and the use of tobacco. *Amer. Statistician*, **8**, 7–13.

Dyke, G.V. and Patterson, H.D. (1952) Analysis of factorial arrangements when the data are proportions. *Biometrics*, **8**, 1–12.

Edwards, A.W.F. (1963) The measure of association in a 2 × 2 table. *J.R. Statist. Soc.*, **A126**, 109–14.

Efron, B. (1975) The efficiency of logistic regression compared to normal discriminant analysis. *J. Amer. Statist. Assoc.*, **70**, 892–8.

Efron, B. (1988) Logistic regression, survival analysis and the Kaplan–Meier curve. *J. Amer. Statist. Assoc.*, **81**, 321–7.

Engel, J. (1983) The analysis of dependent count data. Wageningen thesis.

Engel, J. (1988) Polytomous logistic regression. *Statistica Neerlandica*, **42**, to appear.

Feldstein, M.S. (1966) A binary variable multiple regression method of analysing factors affecting perinatal mortality and other outcomes of pregnancy. *J.R. Statist. Soc.*, **A129**, 61–73.

Feller, W. (1968) *An Introduction to Probability Theory and its Applications*, Vol. 1, 3rd edn, Wiley, New York.

Fidler, V. (1984) Change-over clinical trial with binary data: mixed-model based comparison of tests. *Biometrics*, **40**, 1063–70.

Fienberg, S.E. (1977) *The Analysis of Cross-classified Categorical Data*, MIT Press, Cambridge, Mass.

Finney, D.J. (1952) *Probit Analysis*, 2nd edn, Cambridge University Press.

Finney, D.J. (1964) *Statistical Method in Biological Assay*, 2nd edn, Griffin, London.

Firth, D. (1987) On the efficiency of quasi-likelihood estimation. *Biometrika*, **74**, 233–45.

Fisher, R.A. (1922) On the interpretation of chi square from contingency tables, and the calculation of *p*. *J.R. Statist. Soc.*, **85**, 87–94.

Fisher, R.A. (1935) The logic of inductive inference (with discussion). *J.R. Statist. Soc.*, **98**, 39–54. Reprinted, without discussion, in

Fisher, R.A. (1950) *Contributions to Mathematical Statistics*, Wiley, New York.

Fisher, R.A. (1956) *Statistical Methods and Scientific Inference*, Oliver and Boyd, Edinburgh.

Fix, E. and Hodges, J.L. (1955) Significance probabilities of the Wilcoxon test. *Ann. Math. Statist.*, **26**, 301–12.

Fowlkes, E.B. (1987) Some diagnostics for binary logistic regression via smoothing. *Biometrika*, **74**, 504–16.

Fowlkes, E.B., Freeny, A.E. and Landwehr J.M. (1988) Evaluating logistic models for large contingency tables. *J. Amer. Statist. Assoc.* **83**, 611–22.

Freeman, M.F. and Tukey, J.W. (1950) Transformations related to the angular and the square root. *Ann. Math. Statist.*, **21**, 607–11.

Fuller, W.A. (1987) *Measurement Error Models*, Wiley, New York.

Gart, J.J. (1969) An exact test for comparing matched proportions in crossover designs. *Biometrika*, **56**.

Gart, J.J. (1970) Point and interval estimation of the common odds ratio in the combination of 2 × 2 tables with fixed marginals. *Biometrika*, **57**, 471–5.

Gart, J.J. (1971) The comparison of proportions: a review of significance tests, confidence intervals and adjustments for stratification. *Int. Statist. Rev.*, **39**, 148–89; errata in **40**, 221.

Gart, J.J. (1985) Approximate tests and interval estimation of the common relative risk in the combination of 2 × 2 tables. *Biometrika*, **72**, 873–7.

Gart, J.J., Pettigrew, H.M. and Thomas, P.G. (1985) The effect of bias, variance estimation, skewness and kurtosis of the empirical logit on weighted least squares analyses. *Biometrika*, **72**, 179–90.

Godambe, V.P. and Heyde, C.C. (1987) Quasi-likelihood and optimal estimation. *Int. Statist. Rev.*, **55**, 231–44.

Goodman, L.A. (1963) On Plackett's test for contingency table interactions. *J.R. Statist. Soc.*, **B25**, 179–88.

Goodman, L.A. (1985) The analysis of cross-classified data having ordered and/or unordered categories: association models, correlation models, and asymmetry models for contingency tables with or without missing entries (with discussion). *Ann. Statist.*, **13**, 10–69.

Goodman, L.A. (1986) Some useful extensions of the usual correspondence approach and the usual log-linear models approach in the analysis of contingency tables (with discussion). *Int. Statist. Rev.*, **54**, 243–309.

Goodman, L.A. and Kruskal, W.H. (1954) Measures of association for cross classifications. *J. Amer. Statist. Assoc.*, **49**, 732–64.

Goodman, L.A. and Kruskal, W.H. (1959) Measures of association for cross classifications. II. Further discussion and references. *J. Amer. Statist. Assoc.*, **54**, 123–63.

Goodman, L.A. and Kruskal, W.H. (1963) Measures of association for cross classifications. III. Approximate sampling theory. *J. Amer. Statist. Assoc.*, **58**, 310–64.

Gordon, T. and Foss, B.M. (1966) The role of stimulation in the delay of onset of crying in the new-born infant. *J. Exp. Psychol.*, **16**, 79–81.

Green, P.J. (1984) Iteratively reweighted least squares for maximum likelihood estimation and some robust and resistant alternatives (with discussion). *J.R. Statist. Soc.*, **B46**, 149–92.

Greenacre, M.J. (1984) *Theory and Application of Correspondence Analysis*, Academic Press, New York.

Grizzle, J.E. (1961) A new method of testing hypotheses and estimating parameters for the logistic model. *Biometrics*, **17**, 372–85.

Guererro, M. and Johnson, R.A. (1982) Use of the Box-Cox transformation with binary response models. *Biometrika*, **65**, 309–14.

Haldane, J.B.S. (1955) The estimation and significance of the logarithm of a ratio of frequencies. *Ann. Hum. Genetics*, **20**, 309–11.

Haldane, J.B.S. and Smith, C.A.B. (1948) A simple exact test for birth-order effect. *Ann. Eugenics*, **14**, 117–24.

Hastie, T. and Tibshirani, R. (1986) Generalized additive models (with discussion). *Statistical Science*, **1**, 297–318.

Hastie, T. and Tibshirani, R. (1987a) Generalized additive models: some applications. *J. Amer. Statist. Assoc.*, **82**, 371–86.

Hastie, T. and Tibshirani, R. (1987b) Non-parametric, logistic and proportional odds regression. *Appl. Statist.*, **38**, 260–76.

Hauck, W.W. (1979) The large-sample variance of the Mantel–Haenszel estimator of a common odds ratio. *Biometrics*, **35**, 817–19.

Hauck, W.W. (1984) A comparative study of conditional maximum likelihood estimation of a common odds ratio. *Biometrics*, **35**, 817–19.

Hauck, W.W. and Donner, A. (1988) The asymptotic relative efficiency of the Mantel–Haenszel estimator in the increasing-number-of-strata case. *Biometrics*, **44**, 379–84.

Hewlett, P.S. and Plackett, R.L. (1964) A unified theory for quantal responses to mixtures of drugs: competitive action. *Biometrics*, **20**, 566–75.

Hitchcock, S.E. (1966) Tests of hypotheses about the parameters of the logistic distribution. *Biometrika*, **53**, 535–44.

Hodges, J.L. (1958) Fitting the logistic by maximum likelihood. *Biometrics*, **14**, 453–61.

Hutchison, D. (1985) Ordinal variable regression using the McCullagh (proportional odds) model. *GLIM Newsletter*, no. 9. p. 9.

Imrey, P.B., Koch, G.G. and Stokes, M.E. (1981, 1982) Categorical data analysis: some reflections on the log linear model and logistic regression; Part I: Historical and methodological overview. *Int. Statist. Rev.*, **49**, 265–83. Part II: Data analysis. *Int. Statist. Rev.*, **50**, 35–63.

Jewell, N.P. (1984) Small-sample bias of point estimators of the odds ratio from matched sets. *Biometrics*, **40**, 421–36.

Jørgensen, B. (1983) Maximum likelihood estimation and large-sample inference for generalized linear and nonlinear regression models. *Biometrika*, **70**, 19–28.

Kalbfleisch, J.D. and Prentice, R.L. (1980) *The Statistical Analysis of Failure Time Data*, Wiley, New York.

Kay, R. and Little, S. (1987) Transformations of the explanatory variables in the logistic regression model for binary data. *Biometrika*, **74**, 495–501.

Kedem, B. (1980) *Binary Time Series*, Marcel Dekker, New York.

Kendall, M.G. and Stuart, A. (1963) *The Advanced Theory of Statistics*, vol. 1, 2nd edn, Griffin, London.

Kendall, M.G. and Stuart, A. (1967) *The Advanced Theory of Statistics*, vol. 2, 2nd edn, Griffin, London.

Kenward, M.G. and Jones, B. (1987) A log linear model for binary and cross-over data. *Appl. Statist.*, **36**, 192–204.

Kenward, M.G. and Jones, B. (1989) *Cross-over Trials*, Chapman and Hall, London.

Kupper, L.L., Portier, C., Hogan, M.D. and Yamamoto, E. (1986) The impact of litter effects on dose-response modelling in teratology. *Biometrics*, **42**, 85–98.

Landwehr, J.M., Pregibon, D. and Shoemaker, A.C. (1984) Graphical methods for assessing logistic regression models. *J. Amer. Statist. Assoc.*, **79**, 61–71.

Lauritzen, S.L. and Wermuth, N. (1989) Graphical models for association between variables, some of which are qualitative and some quantitative. *Ann. Statist.*, to appear.

Lebart, L., Morineau, A. and Warwick, K.M. (1984) *Multivariate Descriptive Statistical Analysis*, Wiley, New York.

Lehmann, E.L. (1986) *Testing Statistical Hypotheses*, 2nd edn, Wiley, New York.

Lewis, B.N. (1962) On the analysis of interaction in multidimensional contingency tables. *J.R. Statist. Soc.*, **A125**, 88–117.

Liang, K.Y. (1985) Odds ratio inference with dependent data. *Biometrika*, **72**, 678–82.

Lindley, D.V. (1964) The Bayesian analysis of contingency tables. *Ann. Math. Statist.*, **35**, 1622–43.

Lombard, H.L. and Doering, C.R. (1947) Treatment of the four-fold table by partial correlation as it relates to Public Health problems. *Biometrics*, **3**, 123–8.

Lubin, J.H. (1981) An empirical evaluation of the use of conditional and unconditional likelihoods for case-control data. *Biometrika*, **68**, 567–71.

McCullagh, P. (1980) Regression models for ordinal data (with discussion). *J.R. Statist. Soc.*, **B42**, 109–42.

McCullagh, P. (1983) Quasi-likelihood functions. *Ann. Statist.*, **11**, 59–67.

McCullagh, P. (1986) The conditional distribution of goodness-of-fit statistics for discrete data. *J. Amer. Statist. Assoc.*, **81**, 104–7.

McCullagh, P. and Nelder, J.A. (1983) *Generalized Linear Models*, Chapman and Hall, London.

McNemar, Q. (1947) Note on the sampling error of the differences between correlated proportions or percentages. *Psychometrika*, **12**, 153–7.

Mantel, N. and Haenszel, W. (1959) Statistical aspects of the analysis of data from retrospective studies of disease. *J. Nat. Cancer Inst.*, **22**, 719–48.

Mantel, N. and Hankey, W. (1975) The odds ratio of a 2×2 contingency table. *Amer. Statistician*, **29**, 143–5.

Mariotto, A. (1988) An empirical Bayes analysis of a collection of 2 × 2 tables. Unpublished chapter of Imperial College thesis.

Maritz, J.S. (1989). *Empirical Bayes Methods*, 2nd edn, Chapman and Hall, London.

Maxwell, A.E. (1961) *Analysing Qualitative Data*, Chapman and Hall, London.

Minkin, S. (1987) Optimal designs for binary data. *J. Amer. Statist. Assoc.*, **82**, 1098–103.

Molenaar, I.W. (1983) Some improved diagnostics for failure of the Rasch model. *Psychometrika*, **48**, 49–72.

Montgomery, M.R., Richards, T. and Braun, H.I. (1986) Child health, breast-feeding and survival in Malaysia: a random-effects logit approach. *J. Amer. Statist. Assoc.*, **81**, 297–309.

Naylor, A.F. (1964) Comparisons of regression constants fitted by maximum likelihood to four common transformations of binomial data. *Ann. Hum. Genet.*, **27**, 241–6.

Olkin, I. and Tate, R.F. (1961) Multivariate correlation models with mixed discrete and continuous variables. *Ann. Math. Statist.*, **32**, 448–65.

Page, E.S. (1955) A test for a change in a parameter occurring at an unknown point. *Biometrika*, **42**, 523–7.

Palmgren, J. (1981) The Fisher information matrix for log linear models arguing conditionally on observed explanatory variables. *Biometrika*, **68**, 563–6.

Palmgren, J. (1987) Precision of double sampling estimators for comparing two probabilities. *Biometrika*, **74**, 687–94.

Palmgren, J. and Ekholm, A. (1987) Exponential family nonlinear models for categorical data with errors of observations. *Appl. Stoch. Models Data Anal.*, **3**, 111–24.

Pearson, E.S. (1947) The choice of statistical tests illustrated on the interpretation of data classed in a 2 × 2 table. *Biometrika*, **34**, 139–67.

Pearson, E.S. and Hartley, H.O. (1966) *Biometrika Table for Statisticians*, 3rd edn, Cambridge University Press.

Pearson, K. (1900) On a criterion that a given system of deviations from the probable in the case of a correlated system of variables is such that it can be reasonably supposed to have arisen from random samples. *Phil. Mag.*, **50**, 157–75.

Pierce, D.A. and Schafer, D.W. (1986) Residuals in generalized linear models. *J. Amer. Statist. Assoc.*, **81**, 977–86.

Pike, M.C., Hill, A.P. and Smith, P.G. (1980) Bias and efficiency in logistic analysis of stratified case-control studies. *Int. J. Epidemiol.*, **9**, 89–95.

Plackett, R.L. (1962) A note on interactions in contingency tables. *J.R. Statist. Soc.*, **B24**, 162–6.

Plackett, R.L. (1981) *The Analysis of Categorical Data*, 2nd edn, Griffin, London.

Plackett, R.L. and Hewlett, P.S. (1967) A comparison of two approaches to the construction of models for quantal responses to mixtures of drugs. *Biometrics*, **23**, 27–44.

Pregibon, D. (1981) Logistic regression diagnostics. *Ann. Statist.*, **9**, 705–24.

Pregibon, D. (1982) Score tests in GLIM with applications. In *Lecture Notes in statistics*, **14**, GLIM. 82: *Proc. of Int. Conf. on Generalized Linear Models* (ed. R. Gilchrist), Springer-Verlag, New York.

Prentice, R.L. (1986) Binary regression using an extended beta-binomial distribution, with discussion of correlation induced by covariate measurement errors. *J. Amer. Statist. Assoc.*, **81**, 321–7.

Prentice, R.L. and Pyke, R. (1979) Logistic disease incidence models and case-control studies. *Biometrika*, **66**, 403–11.

Rasch, G. (1960) *Probabilistic Models for Some Intelligence and Attainment Tests*, Nielson and Lydiche, Copenhagen.

Ries, P.N. and Smith, H. (1963) The use of chi-square for preference testing in multidimensional problems. *Chemical Engineering Progress*, **59**, 39–43.

Robbins, H.E. (1956) An empirical Bayes approach to statistics. *Proc. 3rd. Berkeley Symp.*, **1**, 157–63.

Robins, J., Breslow, N. and Greenland, S. (1986) Estimators of the Mantel–Haenszel variance consistent in both sparse data and large-strata limiting models. *Biometrics*, **42**, 311–24.

Ruiz, S. (1989) Relation between logistic regression and discriminant analysis. Unpublished chapter of Imperial College thesis.

Schaafsma, W. (1973) Paired comparisons with order-effects. *Ann. Statist.*, **1**, 1027–45.

Sewell, W.H. and Shah, V.P. (1968) Social class, parental encouragement and educational aspirations. *Amer. J. Sociol.*, **73**, 559–72.

Simpson, E.H. (1951) The interpretation of interaction in contingency tables. *J.R. Statist. Soc.*, **B13**, 238–41.

Stefanski, L.A. (1985) The effects of measurement error on parameter estimation. *Biometrika*, **72**, 583–92.

Stefanski, L.A. and Carroll, R.J. (1985) Covariate measurement error in logistic regression. *Ann. Statist.*, **13**, 1335–51.

Stefanski, L.A. and Carroll, R.J. (1987) Conditional scores and optimal scores for generalized linear measurement-error models. *Biometrika*, **74**, 703–16.

Stukel, T.A. (1988) Generalized logistic models. *J. Amer. Statist. Assoc.*, **83**, 426–31.

Tarone, R.E. (1985) On heterogeneity tests based on efficient scores. *Biometrika*, **72**, 91–5.

Tibshirani, R. and Hastie, T. (1987) Local likelihood estimation. *J. Amer. Statist. Assoc.*, **82**, 559–67.

Tocher, K.D. (1950) Extension of the Neyman–Pearson theory of tests to discontinuous variables. *Biometrika*, **37**, 130–44.

Tukey, J.W. (1949) One degree of freedom for non-additivity. *Biometrics*, **5**, 232–42.

Wald, A. (1947) *Sequential Analysis*, Wiley, New York.

Walker, S.H. and Duncan, D.B. (1967) Estimation of the probability of an event as a function of several independent variables. *Biometrika*, **54**, 167–79.

Wang, P.C. (1985) Adding a variable in generalized linear models. *Technometrics*, **27**, 273–6.

Wedderburn, R.W.M. (1974) Quasi-likelihood functions, generalized linear models and the Gauss–Newton method. *Biometrika*, **61**, 439–47.

West, M., Harrison, P.J. and Migon, H.S. (1985) Dynamic generalized linear models and Bayesian forecasting (with discussion). *J. Amer. Statist. Assoc.*, **80**, 73–97.

Williams, D.A. (1975) The analysis of binary responses from toxicological experiments involving reproduction and teratogenecity. *Biometrics*, **31**, 949–52.

Williams, D.A. (1982) Extra-binomial variation in logistic linear models. *Appl. Statist.*, **31**, 144–8.

Williams, D.A. (1987) Generalized linear model diagnostics using the deviance and single case deletions. *Appl. Statist.*, **36**, 181–91.

Woolf, B. (1955) On estimating the relation between blood group and disease. *Ann. Hum. Genetics*, **19**, 251–3.

Wu, C.F.J. (1985) Efficient sequential designs with binary data. *J. Amer. Statist. Assoc.*, **80**, 974–84.

Yates, F. (1934) Contingency tables involving small numbers and the χ^2 test. *Suppl. J.R. Statist. Soc.*, **1**, 217–35.

Yates, F. (1955) The use of transformation and maximum likelihood in the analysis of quantal experiments involving two treatments. *Biometrika*, **42**, 382–403.

Yates, F. (1984) Tests of significance for 2×2 contingency tables (with discussion). *J.R. Statist. Soc.*, **A147**, 426–63.

Author index

Subject index